Alfred Goldsborough Mayor

On the Color and Color-Patterns of Moths and Butterflies

Alfred Goldsborough Mayor

On the Color and Color-Patterns of Moths and Butterflies

ISBN/EAN: 9783743435407

Manufactured in Europe, USA, Canada, Australia, Japa

Cover: Foto ©berggeist007 / pixelio.de

Manufactured and distributed by brebook publishing software
(www.brebook.com)

Alfred Goldsborough Mayor

On the Color and Color-Patterns of Moths and Butterflies

Bulletin of the Museum of Comparative Zoölogy

AT HARVARD COLLEGE.

VOL. XXX. No. 4.

ON THE COLOR AND COLOR-PATTERNS OF MOTHS AND BUTTERFLIES.

BY ALFRED GOLDSBOROUGH MAYER.

WITH TEN PLATES.

CAMBRIDGE, MASS., U. S. A. :
PRINTED FOR THE MUSEUM.
FEBRUARY, 1897.

No. 4.—*On the Color and Color-Patterns of Moths and Butterflies.*[1]

By Alfred Goldsborough Mayer.

This research is an investigation of the general phenomena of Color in Lepidoptera, and also a special account of the Color-Patterns of the Danaoid and Acraeoid Heliconidae, and of the Papilios of Tropical South America, and has been carried out under the direction of my friend and instructor, Dr. Charles B. Davenport; and the work was done in connection with one of the courses given by him in Harvard University in 1894-95.[2] I am indebted to Dr. Davenport not only for suggesting the subject, but also for his kindness in devoting much time to a criticism of the results.

The paper is divided into three parts. Part A contains an account of the general phenomena of color in Lepidoptera; Part B is devoted to a special discussion of the color-variations in the Heliconidae, with special reference to the phenomena of mimicry; and Part C consists of a summary of those results which are believed to be new to science. A Table of Contents is given at the end of the paper.

PART A.

GENERAL PHENOMENA OF COLOR IN LEPIDOPTERA

I. Classification of Colors.

We follow Poulton ('90) in dividing Lepidopterous colors into (1) pigmental and (2) structural.

(1) *Pigmental Colors* are due to the presence of an actual pigment within the scales, and although such colors are very common in the Lepidoptera, it is frequently very difficult to say off-hand whether a given color is due to a pigment or to some structural effect. Coste ('90-'91) and Urech ('93) have, however, given criteria for determining whether a color is due to a pigment or to some other cause. They succeeded, for example, in dissolving out the color in many

---

[1] Contributions from the Zoölogical Laboratory of the Museum of Comparative Zoölogy at Harvard College, E. L. Mark, Director, No. LXXIV.

[2] This paper was written in 1895 essentially as it now stands.

cases, leaving the wing white or colorless. Coste used as solvents a number of strong acids and alkalis; while Urech confined himself to the use of water, hydrochloric acid, and nitric acid. Their results may be conveniently summarized as follows : —

Black according to Urech is a pigmental color, for it may be dissolved out of the wings by means of hydrochloric or nitric acid.

Brown is usually insoluble in water, but is soluble in hydrochloric or nitric acid.

The *red and orange* pigments of the Pieridae, Lycaenidae, Nymphalidae, Zygaenidae, and some Papilios are soluble in water. They are insoluble in water in the Sphingidae, Arctidae, Bombycidae, Saturnidae, and Geometridae.

Yellow pigment is acted upon by reagents in almost the same way as the red and orange, especially if both red and yellow appear upon the same wing. It is soluble in the Pieridae, Lycaenidae, Nymphalidae, Satyridae, and some Papilios, but insoluble in the Sphingidae, Arctidae, Geometridae, and a few Noctuidae.

White is usually a structural color, but can be dissolved out from the wings of the Pieridae by water, being in this case, of course, due to a pigment.

Green pigment can be dissolved out by water in the cases of the Pieridae, Lycaenidae, and Geometridae. In the vast majority of cases, however, it is a structural color.

Violet and blue are almost always due to structural causes. In a few cases, however, as in Smerinthus ocellatus, a blue pigment can be dissolved out.

We see, then, that black, brown, red, orange, and yellow are usually due to pigment, while white, green, violet, and blue are generally due to structural effects.

It is well known that the scales of Lepidoptera are essentially hollow, flattened sacs often inclosing pigment, and Burmeister ('78) arrives at the conclusion, from a study of the scales in various species of Castnia, that the pigment is for the most part attached to the upper layer of the scale-sac, rendering it opaque, while the lower layer receives less pigment and is, in consequence, a little more translucent.

(2) *Structural Colors* owe their origin to the external structure of the scales or wing-membranes and not to the presence of a pigment. They are often caused by diffraction, due to the scales being covered with fine, parallel striae. Some of the most splendid colors in the

animal kingdom are due to this cause; such are the iridescent and opalescent hues of many of the Morphos and Indo-Asiatic Papilios. Very often the scales which display such brilliant colors contain no pigment whatsoever; for if one will merely soak them in alcohol, ether, or water, all color disappears, and the scales become as transparent as glass. This test was devised by Dimmock ('83), who used it upon the brilliantly colored scales of many beetles. It was first discovered by Burgess ('80), and has since been confirmed by Kellogg ('94), that the striae which produce these structural colors are all upon the outer surface of the scale, *i. e.*, the surface which is away from the wing-membrane and exposed to the light. Kellogg ('94) has determined the distance apart of the striae upon the scales of many species of Lepidoptera. It appears, for example, that the striae upon the scales of Danais plexippus are 2μ apart, those upon the transparent scales of Morpho sp. 1.5μ, upon the pigment-bearing scales of Morpho 0.72μ, and upon Callidryas eubule 0.9μ apart. It is very evident, then, that the brilliant coloration of the scales may be due to this fine striation, for the striae upon Rowland's or Rutherfurd's finest gratings are approximately 1.5μ apart, which is about the average distance between the ridges of the scales.

Structural colors are, however, not always due to diffraction; in the case of white, for example, the color is almost invariably due to a reflection of all, or nearly all, the light that impinges upon the scales. As long ago as 1855 Leydig pointed out that the silvery white color seen in the scales of some spiders, such as Salticus and Tegenaria, was due to air contained within them; and more recently Dimmock ('83) has shown that silvery white and milk-white colorations are due to optical effects produced by reflected light. In the silvery white scales, however, such as those of the under surface of the hind wings of Argynnis, there must be a polished reflecting surface toward the observer, for both silvery and milk-white colors appear simply milk-white by reflected light.

(3) *Combination Colors* owe their richness and brilliancy to a combination of structural and pigmental effects. The geranium-red spots upon the hind wings of the Mexican Papilio zenuis Lucas owe their red color to pigment, but over this red there plays, in certain lights, a beautiful pearly iridescence, which, in combination with the red, greatly enhances its charm. Urech ('92) has demonstrated that in the Vanessas there are scales which have chemical coloring matter

and interference colors also. In addition, he points out the interesting case of certain Lycaenidae where the scales exhibit to the eye only interference effects, and yet a pigment can be dissolved out of them by the use of water.

(4) *Quantitative Determination of Pigmental Colors.* I have analyzed the colors of many butterflies by means of the spectroscope, and also by Maxwell's discs. As is well known, Maxwell's discs are colored circular discs of cardboard, perforated at the center and slit along a radius so that two or more of them may be slid over each other, thus exposing different proportions of each. Then by rapidly rotating them the colors become blended, and thus it becomes possible to match any color, and to discover its fundamental constituents. By this means I have determined that the vast majority of the colors found in Lepidoptera are impure; that is to say, they contain a large percentage of black.

For example the white of the upper surface of the wings of the common Pieris rapae consists of: 17% black, 13% emerald-green, 10% lemon-yellow, and 60% white.

Also the so-called "blacks" found in butterflies are rarely jet-black, but, almost always, only deep shades of brown. For instance the deep brown color of the under surface of the wings of Heliconius melpomene consists of 93% black, 3% lemon-yellow, 3.5% of Maxwell's fundamental red (vermilion), and 0.5% of von Bezold's fundamental blue-violet.

The purest color I have met with is the canary-yellow ground color of the wings of Papilio turnus, which seems to consist of white light with the addition of a little yellow.

Other colors all possess considerable black. Thus the glaucous green of Colaenis dido consists of black 29%, vermilion 24%, emerald-green 37%, von Bezold's blue-violet 10%.

The sepia-brown ground color of Cercyonis alope consists of black 71%, vermilion 21.5%, emerald-green 7.5%.

The tawny rufous color of the wings of Mechanitis polymnia, etc., is made up of black 46%, vermilion 40%, lemon-yellow 14%.

The rufous red patch on the upper surface of the fore wings of Heliconius melpomene is made up of black 27%, vermilion 66.5%, lemon-yellow 6.5%.

The yellow of the fore wings of Mechanitis polymnia consists of lemon-yellow 67%, emerald-green 14%, and white 19%.

(5) *Spectrum Analysis of Colors of Lepidoptera.* I have made some spectrum analyses of the light reflected from the wings of various butterflies, by means of a piece of apparatus most kindly suggested for the purpose by Prof. Ogden N. Rood of Columbia College. The arrangement is shown in Figs. 1, 2, Plate 1; Fig. 1 being a perspective view, and Fig. 2 a horizontal section of the apparatus, which consists of a rectangular box, blackened upon the inside, and having a well-fitting cover. A rectangular slit (O) was cut through one of the long sides of the box, near one end, and the other end of the same side was perforated in order to allow the admission of the direct-vision spectroscope (S). Imagine that we wish to examine the yellow spots from a butterfly's wing. All of the yellow spots from the wing are cut out, and pasted upon two pieces of cardboard so as to make two large unbroken patches of color. The pieces of cardboard are then blackened upon all those places where the colored wing was not pasted. One of the cardboards is then suitably mounted upon the back of the box at B; the other is placed upon a vertical support (F), the plane of which is parallel to the back of the box.

The working of the apparatus is as follows: the sunlight enters by the slit (O) and is reflected and diffused three or four times between the pieces of colored wing mounted upon the back (B) of the box, and the vertical support (F). The manner of this reflection and diffusion is shown by the dotted lines of Fig. 2. After undergoing several reflections, the light enters the direct-vision spectroscope (S). The slit of the spectroscope is wide open, and thus the light which enters it may readily be examined. It was found that it was necessary that the light be reflected more than once from the wing before it enters the spectroscope, for the first reflection shows so much white light that it is usually quite impossible to analyze the true color of the wing, the predominant colors being obscured by a continuous spectrum. In general it was found that the colors of the wings are not simple, but compound; that is to say, they are made up of a mixture of several different colors.

For example, the spectrum of the rufous ground color of the upper surface of the wings of Danais plexippus consists of all of the red and yellow of the spectrum and about 75% of the green.

The red spots upon the upper side of the fore wings of Heliconius melpomene also consist of the red and yellow and a very faint, hardly visible, trace of green.

The glaucous green patches on the wings of Colaenis dido are composed mainly of green and yellow, but there is also a faint development of about half of the blue and a still fainter trace of red.

The iridescent blue-green ground color of the upper surface of the wings of Morpho menelaus, viewed in such a way that the light makes an angle of about 20° with the normal to the surface of the wing, gives a spectrum of green and blue about equally developed.

The yellow ground color found on the upper side of the wings of Papilio turnus shows a continuous spectrum, in which the yellow seems to be rather more brilliant than in the normal spectrum of white light.

The sepia-brown ground color of the upper surface of the wings of Cercyonis alope gives a spectrum which lacks only the blue-green and blue.

(6) *Summary of Results.* The researches of Coste ('90–'91) and Urech ('93) have demonstrated that the colors of butterflies and moths may be produced by two causes : by the presence of an actual pigment, or by some structural effect. Some colors are due entirely to pigment, others to structural causes, and still others to a combination of the two.

Black, brown, red, orange, and yellow are invariably due to pigment.

Green is usually due to a structural effect, but in a few cases there is a green pigment present.

White, blue, and violet are almost invariably due to structural causes.

In addition to these facts I have found that most of the colors which are displayed by Lepidoptera contain a surprisingly large percentage of black. Also they are usually not simple colors, but composed of a mixture of several different colors. It is remarkable that Natural Selection, which is generally assumed to have been one of the principal factors in bringing about the wonderful development of colors in Lepidoptera, has not been potent enough to make these colors purer than is the case in existing butterflies.

II. The essential Nature of Pigmental Color in Lepidoptera.

(1) *Pigments of Larvae.* Poulton ('85) showed that the phytophagous larvae of Lepidoptera " owe their colour and markings to

two causes: (1) Pigments derived from their food-plants, chlorophyll and xanthophyll, and probably others; (2) pigments proper to the larvae, or larval tissues made use of because of some (merely incidental) aid which they lend to the colouring, e. g. fat." Poulton concludes that all green coloration is due to chlorophyll, and that nearly all yellows are due to xanthophyll. All other colors, including black and white and some yellows, are due to pigments proper to the larvae themselves.

Later, in 1893, Poulton proved that the larvae of Tryphaena pronuba could transform both etiolin and chlorophyll into a larval coloring matter, which may be either green or brown. It thus appears that some brown pigments are derived from food, and are merely modified plant pigments. Green larvae have green blood, and this color is due to chlorophyll in solution. It is remarkable that this chlorophyll solution is stable under the prolonged action of light, and in this respect is different from any other known solution of chlorophyll. It is worthy of note, further, that the spectrum of this green blood shows a great resemblance to that of chlorophyll. "In fact the two spectra are far nearer to one another than the ordinary spectrum of chlorophyll in alcoholic solution, is to the unaltered chlorophyll of leaves."

(2) *Pigments of Imagines.* In 1891, Urech showed that the similarity between the color of the urine of butterflies and the principal color of their scales is so close that it cannot be considered as accidental, but rather must be regarded as physiological. Urech compares in a table the color of the urine and that of the scales of 29 species of Lepidoptera. In all but two species the resemblance is very close.[1]

Urech further shows that the color of the urine (and the corresponding color of the scales) is not dependent upon the kind of food, for one and the same food plant may be differently digested in different groups of Lepidoptera. Thus he compares the behavior of a Vanessa with that of one of the Microlepidoptera (leaf-rollers). Both of these feed upon the nettle (Urtica). In the larva of the Vanessa the contents of the stomach are intensely green, but become red in the pupa. In the case of the leaf-roller the contents of the stomach are never markedly green and become insipid in color during the pupal stage.

[1] Likewise, Hopkins ('94) has shown that in the Pieridae the urine is tinged by a yellow substance having exactly the color of the wings.

Poulton has shown that the reddish fluid voided by the Vanessas immediately after emergence from the chrysalis contains uric acid, and Hopkins ('94) says that when the yellow Pieridae emerge, they often void from the rectum a large quantity of uric acid. It should be borne in mind however, as Urech himself suggests, that the pigment found within the wings may not be identical in chemical composition with the similarly colored fluid from the alimentary tract.

Hopkins ('89, '91, '94, '96) has discovered that the white pigment found in the scales of Pieridae is uric acid, and that the red and yellow pigments of the Pieridae are due to derivatives of uric acid. He also says, " these uric acid derivatives used in ornamentation, are apparently confined to the Pieridae alone among butterflies." Hence when a Pierid mimics an insect of another family, the pigments in the two cases are chemically quite distinct. This is well seen in the genera Leptalis (Pieridae) and Mechanitis (Danaidae).

In addition to this, Griffiths ('92) finds that the green pigment found in Papilio, Parthenos, Hesperia, Limenitis, Larentia, Ino, and Halias is a derivative of uric acid, to which he gives the name of "Lepidopteric acid" and assigns the empyrical formula $C_{11} H_{16} Az_2 N_x O_{10}$.

In a paper published in 1896 in the Bulletin of the Museum of Comparative Zoölogy at Harvard College, Vol. 29, I have shown, p. 226-230, that the pigments of the scales of Lepidoptera are derived by various chemical processes from the blood, or haemolymph, of the pupa, and that the haemolymph is a proteid substance containing egg-albumen, globulin, fibrin, xanthophyll, orthophosphoric acid, iron, potassium, and sodium.

III. DEVELOPMENT OF THE VARIOUS COLORS IN THE PUPAL WINGS.

A few researches have been carried out upon this interesting topic, but as the literature is scattered and has never been brought together, it will perhaps not be amiss to present a brief résumé of the principal facts which have been already ascertained.

(1) *Historical Account of previous Researches.* In 1889 Schäffer ('89) discussed the question of the order and time of appearance of the colors in the pupal wings of several of the Vanessas. Unfortunately he apparently did not make his obser-

vations at sufficiently close intervals of time, and was, therefore, led into some misstatements, which have been corrected by van Bemmelen ('89) and Urech ('91).

Van Bemmelen carried out an elaborate research upon the development of the various spots and colors upon the wings of Pyrameis cardui, Vanessa urticae, V. io, Pieris brassicae, and a few other forms. He discusses in detail the time and manner of appearance of all of the different spots upon the wing. Into these details we shall not follow him, but shall merely present his general conclusions regarding the development of the various colors. In Pieris brassicae it appears that during the first days of the pupal stage the wings are colorless and transparent; after a few days, however, the fore wings become opaque, and white; later the hind wing, also, goes through the same changes. The wings then remain unaltered until about two days before the butterfly issues. Then, very suddenly, the black spots and the yellow ground tone of the under sides appear. White is thus the primary color; black and yellow secondary. The first color to make its appearance in the case of Pyrameis cardui is a brown-yellow ground color, which may be observed in pupae four days old. The hind wings are at this time somewhat darker than the fore wings. The color then changes from darker brown to cinnamon-brown. The black spots appear later upon this delicate reddish brown ground color. The three fused spots which form the whitish band in the middle of the front edge of the fore wing appear during the last days of development, just before the completion of the final color-pattern.

Both van Bemmelen and Urech have shown that in Vanessa urticae the order of appearance of the various colors is the same as in Pyrameis cardui. The first color to appear in Vanessa urticae is a faint reddish tinge; this deepens and forms the ground color, and later the black spots appear upon it.

Urech ('91) has made a careful study of the development of the colors upon the pupal wings of Vanessa io. The wings are at first wholly white. Then in a restricted area of this white is noticed the appearance of a yellow, which forms the yellow of the mature wings. Almost contemporaneous with the development of the yellow comes the red, which appears in another part of the primitively white field, and gradually deepens in color until it forms the brownish red ground color of the adult wings. Still later another portion of the primitive white changes into the black of the mature wing. The under side of the mature wings of Vanessa io is mainly uniform black, and in this case also this color develops from the white at a very rapid rate, near the end of the pupal stage. This development of the black directly upon the white areas is quite remarkable in Vanessa io, and very different from that of both Vanessa urticae and Pyrameis cardui, where the black spots develop upon a field already tinged with red. Urech points out the fact, that some of the white spots seen in the mature wings of the Vanessas represent the " primitive white " of the pupal wings.

Finally, the latest paper upon the subject of the development of color in the pupa is that of Haase ('93), who has examined the pupae of a number of Papilios (e. g., philenor, machaon, asterias, turnus, and podalirius), and finds that during early pupal life the wings are as transparent as glass; after a time, however, they change to an impure white, which soon becomes yellowish, and then the various colors which are destined to adorn the mature wings begin to appear.

If we are to learn much of fundamental import concerning the phylogeny of color in Lepidoptera, the researches should be carried out upon the lower moths, and not upon such highly specialized forms of Rhopalocera as the Vanessae.

In my paper on Wing scales, etc. (Mayer, '96, p. 232), I have come to the conclusion that dull ocher-yellow and drabs are, phylogenetically speaking, the oldest pigmental colors in the Lepidoptera. The more brilliant colors, such as bright yellows, reds, and pigmental greens, are derived by complex chemical processes, and are, phylogenetically speaking, of recent appearance.

I have made a study of the development of the colors and pattern in the wings of Callosamia promethea Linn. and of Danais plexippus Fab.

(2) *Development of Color in the Pupal Wings of Callosamia promethea.* The cocoons of Callosamia promethea are very abundant during the winter months, when they may be found hanging to the stems of the food plants of the larvae. The pupal wings remain perfectly transparent all through the winter, until about ten days before the time when the moth is destined to issue ; they then become opaque white. An examination of the wings at this period shows that the scales are perfectly formed (Fig. 25, Plate **3**), except for the

lack of pigment, which is developed later. If one treats the scales at this stage with oil of cedar-wood or clove oil, they become practically invisible under the microscope, thus demonstrating that there is no pigment within them. Fig. 26, Plate 3, gives the appearance presented by a scale taken from the light drab-colored margin of the mature wing. This is about the lightest area upon the wing, except the white spots; but it will be seen that this scale is much darker in appearance than the unpigmented one shown in Fig. 25. The white or unpigmented condition of the wing lasts for about four days. The wings then become uniformly tinged with an impure yellow or light drab, and very soon after this the colors begin to make their appearance. They first appear upon the lower surface of the wings. Fig. 28, Plate 3, represents the under surface of the fore wing of a female in a very early stage of color development; in fact the upper surface shows, as yet, no trace of the colors. It will be seen that a few dark red streaks have appeared near the central portion of the wing, and it is worthy of note that these occupy the *interspaces between the nervures.* The ocellus near the apex of the wing appears faintly outlined upon its background of impure yellow.

Fig. 27, Plate 3, represents the under side of a hind wing of a male in about the same stage as Fig. 28. Here, again, the red color occupies the *interspaces*, and indeed it is only later that the nervures become clouded over by it.

Figs. 29 and 30, Plate 3, represent, respectively, the under and upper sides of the fore wing of a male about five hours after the first appearance of the colors. Upon the upper side (Fig. 30) we see two gray streaks near the base of the wing and a light cinnamon-brown color extending from the lower edge toward the middle of the wing. The ocellus near the apex is now quite apparent, but still faint in color. On the under surface (Fig. 29) the red markings have developed to a much greater extent than in Fig. 28. The outermost of the two white spots which occupy the center of this red area becomes the white central spot of the mature wing; the inner-most one is soon obliterated owing to its becoming clouded over with red.

Figs. 37 and 36 represent respectively the upper surface of the fore wing and the lower surface of the hind wing of a female, slightly more advanced than in Fig. 30. Fig. 31 represents a male and Fig. 38 a female about twelve hours after the first appearance of the color. It is remarkable that in this stage the male and female wings are quite similar in general appearance, except that the ground color of the male is now a dusky gray, while that of the female is a cinnamon-brown.

From this time onward, however, the wings of the two sexes begin to differ more and more in appearance, for the ground color of the male becomes deep black, while that of the female remains cinnamon-brown. This change is well exhibited by Figs. 32 and 39, Plate **3**, which give the appearance of the upper surfaces of the male and female wings respectively at about twenty hours after the first appearance of the colors. Fig. 33 represents the hind wing of the same male whose fore wing is shown in Fig. 32. Figs. 34, 35, 40, and 41 give the appearance of the pupal wings just before emergence, when the colors are completely formed.

To summarize; Figs. 27, 29, 33, and 35 give successive stages in the development of color in the male; and Figs. 28, 36–41 give similar stages for the female. It becomes evident, from a comparison of these successive developmental stages, that the colors appear first upon the central portions of the wings, and that the outer and costal edges of the wings and the nervures are the last parts to acquire the mature coloration.

It is worthy of remark that the color-pattern of the mature male Callosamia promethea is quite a departure from the type of coloration which is commonly found among the Saturnidae. The *female*, however, conforms very well to the general pattern of the other species of the family. It is quite evident that the deep black coloration of the male is, phylogenetically speaking, a new acquisition, and that the coloration of the female represents the less differentiated and therefore, more primitive type.

It is interesting in connection with these facts to observe that the color-patterns of both male and female develop in almost identical ways up to the twelfth hour after the first appearance of the color; that then, however, the grayish ground color of the male wings begins to deepen into the characteristic jet black of the adult, while the light cinnamon ground color of the female merely becomes slightly darker as the wings mature.

(3) *Development of Color in the Pupal Wings of Danais plexippus.* Figs. 42–45, Plate **3**, are intended to illustrate four stages in the development of color in the pupal fore wings of Danais plexippus. The pupal stage of this species is of brief duration, last-

ing from one to two weeks only, according to the temperature to which the chrysalis is exposed. For the first few days the wings are perfectly transparent, but about five days before the butterfly issues they become pure white. An examination of the scales at this period shows that they are completely formed and merely lack pigment. In about 48 hours after this (see Fig. 42) the ground color of the wings changes to a dirty yellow. It is interesting to note that the white spots which adorn the mature wings remain pure white. Fig. 43 illustrates the next stage, where the black has begun to appear in the region beyond the cell. The nervures themselves, however, remain white. Fig. 44 shows a still later condition, where the dirty yellow ground color has deepened into rufous, and the black has deepened and increased in area and has also begun to appear along the edges of the nervures. In Fig. 45 the black has finally suffused the nervures, the base of the wing and the submedian nervure being the only parts that still remain dull yellow. It is apparent that in Danais plexippus, as in Callosamia promethea, the central areas of the wings are the first to exhibit the mature colors, and that the nervures and costal edges of the wings are the last to be suffused.

IV. The Laws which govern the Color-Patterns of Butterflies and Moths.

(1) *Historical Account of previous Researches.* The earliest paper upon this subject is by Higgins ('68). He came to the conclusion, that "the simplest type of color presents itself in the plain uniform tint exhibited when the scales are all exactly alike." He also thought it probable that "the scales growing on the membrane upon or near the veins would be distinguished from the scales growing on other parts of the membrane by a freer development of pigmentary matter, and that in this manner would arise a kind of primary or fundamental color-pattern, namely, a pale ground with darker linear markings following the course of the veins, e. g. Pieris crataegi." He also attempted to explain the formation of eye-spots by assuming that crescent-shaped markings migrate outwards from the sides of the nervures and meet so as to inclose a space.

It is, however, untrue that there is a freer development of pigment within the scales lying upon the nervures; in fact, the reverse is the case, as we have seen, in both Danais plexippus and Callosamia promethea. Higgins's explanation of the formation of eyespots is also fallacious.

Darwin ('71, Vol. 2, p. 133) published four excellent figures from a drawing by Trimen, illustrating two simple ways in which *eyespots* are actually formed, both diametrically opposed to Higgins's hypothesis. Darwin says that in the South African butterfly, Cyllo leda, "in some specimens, large spaces on the upper surface of the wings are coloured black, and include irregular white marks, and from this state a complete gradation can be traced into a tolerably perfect ocellus, and this results from the contraction of the irregular blotches of colour. In another series of specimens a gradation can be followed from excessively minute white dots, surrounded by a scarcely visible black line, into perfectly symmetrical and large ocelli" with several rings.

Scudder ('88-'89) and, afterwards, Bateson ('94) have shown that the ordinary eye-spots, such as those found in Morpho and the Satyridae, are invariably placed in the interspaces between the longitudinal veins of the wings, and also that they are often found repeated upon homologous places of both pairs of wings. Bateson says that ocelli are often seen upon both surfaces of the wing, the centers of the upper and lower ocelli coinciding. In the majority of cases, however, the upper and lower ocelli, although coincident, have quite different colors. The simpler sort of ocelli, such as those seen in the Satyridae or in Morpho, have their centers on the line of the foldmarks or creases of the wing. It sometimes happens that these creases seem to begin from the center of an ocellus. As these creases commonly run midway between two nervures, it usually results that the center of the eye-spot is exactly half way between two nervures. The large eye-spots of Parnassius apollo are an exception to this rule. In some Morphos, Satyridae, etc., in cell 1^b of the hind wing there are often two creases and two eye-spots, one for each crease; but if there be only one eye-spot present, its center does not correspond with the middle of the cell, "but is exactly upon the anterior of the two creases." I have observed the same law for the white marginal spots in cell 1^b in Ceratinia vallonia, C. fimbria, and Mechanitis polymnia.

In 1889 Scudder, in his work upon the Butterflies of New

England, called attention to the following facts: the transverse series of dark spots so often seen in the body of the wings of Lepidoptera are invariably placed *in the interspaces between* the longitudinal veins, never upon the veins themselves, excepting only in rare instances, where the spots occur at the extreme margin. He also pointed out that in many types of moths all differentiation in coloring has been greatly retarded, so far as the hind wings are concerned, by their almost universal concealment by day beneath the overlapping front wings. In these cases " the simplest departure from uniformity consists of a deepening of the tint next the outer margin of the wing." It is but a step from this condition to a band of dark color or a row of spots parallel with the margin. This explains why the transverse style of markings, for the hind wings at least, is so common. Scudder showed that "the number of instances, in butterflies, in which similar markings appear in the same areas of the two wings, and in the same relative position in these areas, is far too common to be a mere coincidence. It is most readily traced in the disposition of the ocelli, which are very apt to be similar in size and perfection, and to be situated between the same branches of homologous veins."

(2) *Laws of Color-Patterns.* As a result of my own study of the wings of moths and butterflies, I am prepared to propose the following additional laws of color-patterns. (*a*) Any spot found upon the wings of a moth or butterfly *tends to be bilaterally symmetrical both as regards form and color,* the axis of symmetry being a line passing through the center of the interspace in which the spot is found, and parallel to the direction of the longitudinal nervures. For example, in Figs. 6 and 7, Plate **2**, each spot is bilaterally symmetrical about the axis IIII. The same law holds for the spots represented in Figs. 8–14 and 16.

(*b*) Spots tend to appear *not in one interspace only,* but as a row occupying homologous places in successive interspaces. Indeed we almost always find similar spots arranged in linear series, each similar in shape and color to the others and occupying the center of its interspace. The rows of spots represented in Figs. 8–14 and 16 will suffice to illustrate this law

It is interesting to notice that bands of color are often made by the fusion of a row of adjacent spots; and, conversely, chains of spots are often formed by the breaking up of bands, leaving a row of spots occupying the interspaces. Many instances of this

are to be seen in certain specimens of various species of the
Heliconidae. For example, in Heliconius enerate (Fig. 58, Plate
4) I have observed that certain specimens show a row of distinct
spots in place of the, usually entire, band which crosses the middle
of the hind wing. In fact, the vast majority of bands can be
analyzed into a series of similar elements, each element occupying an
interspace. Thus, in Plate 2, Fig. 17, which represents a wing of
Saturnia spini, the band seen crossing the wing parallel with its
margin is made up of a series of fused crescents, each crescent
occupying an interspace.

If, on the other hand, this band were to break away from the
nervures, the result would be a series of crescent-shaped spots each
occupying the center of an interspace. It is very interesting to
observe the manner in which bands degenerate and disappear.
Numerous opportunities for doing this may be had among the Heliconidae. In some species, as in Melinaea parallelis, hardly any two
specimens are alike in the condition of the black band across the
middle of the hind wings. *The most common method of disappearance is a shrinking away of the band at one end.* This is well illustrated in Figs. 84–87, Plate 7, which represent a sort of "Mercator's
Projection" of the wings of Mechanitis isthmia (for explanation of the
plan of projection see page 207.) Fig. 84 represents a male, showing
a well-marked band of hardly separated spots extending across the
middle of the hind wing. Fig. 87 shows a female in which the
spots are thinner and more crescentic and the separations much
more marked. Fig. 85 is also drawn from a female, in which it will
be seen that the band has shrunk away leaving only a portion of it
at the right, and in Fig. 86, which represents another female specimen, only one faint spot is left.

It is very common to find bands shrinking away at *one* end.
Sometimes, however, they shrink away at *both* ends, and very often
they break up into a row of spots, which may then contract into the
centers of their interspaces and finally disappear. It is worthy of
note that it is *very* rare to find a band breaking at the middle of its
length and each half receding from the other. Such a case is, however, shown by Melinaea parallelis (see Fig. 82, Plate 7), where one
sometimes finds specimens in which the black band across the middle
of the hind wings is complete and unbroken; whereas in other
specimens, as in Fig. 82, it is partially broken in the middle, and in
still others the break has become a wide gap by the drawing away of
the halves of the band from each other.

We see, then, that it is very common to find bands shrinking away from either end, but very rare to find them broken in the middle region. This, however, is only a special case of the law enunciated by Bateson ('94), that the ends of a linear series are more variable than the middle. Almost any row of spots also exhibits the same law, in that the spots occupying the middle portions of the row are similar one to another, while those at the ends of the series depart more or less from the type. (See Figs. 10-13, Plate **2**.)

The position of spots which are situated near the edge of the wing is largely controlled by the wing-folds or creases. In Meli-naea egina (Fig. 96, Plate **8**) there is a row of white spots near the outer edges of the wings, and each of these spots is cut in two by a narrow black line which extends along the wing-fold. Also in Cera-tinia vallonia (Fig. 81, Plate **7**) and in many other forms of the Danaoid Heliconidae one often finds two creases in a cell, and in this case there are two marginal spots, one on each crease. In many other cases, however, the marginal spots are double in each cell, although there is but a single wing-fold; the spots in these cases are situated at some distance on either side of the fold. (See Figs. 95, 96, Plate **8**.) Another very common condition is exem-plified in Fig. 83, Plate **7**, where there is a single marginal spot situated upon the wing-fold in each cell.

(3) *Detailed Discussion of the Laws of Color-Patterns.* Figs. 6-14 and 16, Plate **2**, are taken from special cases which serve to illustrate the two chief laws of color-pattern, *i. e.*, that spots tend to be bilaterally symmetrical about an axis (HH, Figs. 6, 7) passing through the center of the cell parallel with the nervures; and also, that spots of similar shape and color tend to be repeated in a row of adjacent cells.

In Fig. 7 the spots are separated in the middle, but still incline outward symmetrically from the center; indeed, instances of double spots are very common. In such cases, however, each half spot is a reflection of its mate on the other side of the axis passing through the center of the cell.

Fig. 8 represents various eye-spots found in the Morphos, and will serve to illustrate the laws of eye-spots which have been enunci-ated by Scudder ('89) and Bateson ('94). These spots occupy the center of the cells in which they are found. In cell H, for example, is a large eye-spot with a crescent in its center, and it will be observed that this crescent follows the general law and is bilaterally symmetrical about the usual axis.[1]

Fig. 9 shows the law of repetition of some very complex spots, each being bilaterally symmetrical. It is found in Parthenos gambrisius. Figs. 10 and 11 represent Ornithoptera urvilliana and O. priamus respectively. In Fig. 10 we see an instance of a spot within a spot, and in Fig. 11 an even more complex case, for here there are three systems of spots one within another.

Fig. 12 represents the marginal markings found in Hestia jasonia and Fig. 13 Hestia leuconoe var. clara. These two examples are intended to illustrate the fact, that, although the markings are situated *upon* the nervures, they are bilaterally symmetrical not *about the nervures* as axes, but about the usual axis passing midway between the nervures. In Fig. 12 it will be seen that the two curved markings situated upon nervures 1^b and 2, and projecting into cell 1^c, are bilaterally symmetrical only in reference to the axis through the middle of the cell.

In allied species the spot situated upon nervure 1^b is often absent. The system of markings is therefore undergoing degeneration at this end (cf. Fig. 13, cell 1^c). The curved mark upon nervure 5 (Fig. 12) projecting into cell V is plainly symmetrical with respect to its fellow in the opposite side of cell V, and not with its near companion which projects into cell IV. The same is also true in the case of the spots in cell VI.

In Fig. 13 the spots appear at the first glance to be bilaterally symmetrical about both nervures and centers of cells, but in cell IV the marking situated on nervure 4 does not quite reach to the center, and it is interesting to observe that its fellow on nervure 5 also falls short of reaching the center and is therefore symmetrical with respect to the other curved spot in cell IV. This case also furnishes an instance of a break in the middle of a linear series.

Fig. 14 is taken from the under surface of the hind wing of Papilio emalthion. It serves to illustrate the fusion of two originally separate rows of spots. In this case the crescent-shaped spots above have fused with the rectangular ones below, so as to inclose a portion of the ground color of the wing. Sometimes two rows of

---

[1] A very beautiful exception (Fig. 19, Plate 2) to this rule for the crescents found in eye-spots is seen in the under surface of the fore wing of Missanga patima Moore. It will be noticed that the large black crescent found in this beautiful eye-spot is 90° away from its usual position. This is the only exception of the sort known to me.

spots of different colors fuse, giving a chain of spots which are of one color above and another below.

In Fig. 16 the spots composing the row BB are blue (dark) above, and red (light) below. It will be observed that the color is bilaterally symmetrical, as usual, about the axis through the middle of the cell. Such bicolored spots are often due to a simple fusion, as before stated; but sometimes they may, perhaps, be intrinsically bicolored.

Fig. 15 is a beautiful instance of an exception to the general rule that spots are bilateral about the axis through the center of the cell. It is taken from Ornithoptera trojana Staudinger.[1] The light spots represented near the outer edge of the wing are of a brilliant iridescent green. It is evident that they are distinctly bilateral with respect to the *nervures*; especially is this true of the pair adjacent to nervure 1. Ornithoptera brookiana Wallace illustrates another exception, though in a less marked degree.[2] Other allied species of Ornithoptera, however, would seem to show that these apparent exceptions may have been derived from forms which exhibited two spots in each cell and followed the usual rule. These are the only instances of such exceptions known to me. I do not doubt, however, that further study would reveal others.

In Fig. 17 an example is given of the peculiar kind of eye-spots found in the Saturnidae. The species from which the figure was taken is Saturnia spini. It will be seen that this so-called eye-spot is quite different in formation from the ocelli of butterflies. It is simply a series of curved cross-bands between nervures, arranged symmetrically on both sides of the cross vein CC. The "eye-spots" upon the wings of Attacus luna and in the genus Telea are also of this sort. True eye-spots, however, similar to those found among the Morphos and Satyridae, occur in moths, as in the apex of the fore wing of Samia cecropia, Callosamia promethea, etc. "False" eye-spots are also found on the wings of butterflies; in Vanessa io, for example, the so-called eye-spot of the fore wing has been shown by Dixey ('90) to be made up of a series of fused spots. It will be remembered that Merrifield ('94, Plate 9, Fig. 4) caused this "ocellus" to break up into its constituents by subjecting the pupa to a temperature of 1° C. The ocellus upon the hind wing of Vanessa io is no doubt a true eye-spot; the only evidence which

[1] See Watkins, '91, Plate 4.

[2] See Hewitson, '56-'76, Vol. 1.

might lead one to infer that the ocellus of the fore wing was of the same character is, that an aberrant form is sometimes found in nature having the "eye-spots" on both fore and hind wings obliterated, thus indicating a *possible* connection between the two (see South, '89).

Fig. 18 is intended to illustrate the process of degeneration occurring in bands. Band BB is represented as breaking down by the rare method of parting in the middle. Example, Melinaea parallelis. Band EE is degenerating at one end; this is a very common method.

Figs. 20–23 represent hypothetical conditions not found in nature; all being contrary to the conditions of the laws which have just been stated.

In Fig. 20 row RR presents three spots for each cell. I believe this has not been found in nature, but I should not be surprised if it were discovered, for it is not contrary to any of the laws.

Row CC, on the other hand, is contrary to the law of bilaterality, the crescents not being bilateral about axes passing through the middle of the interspaces parallel with the longitudinal nervures.

Fig. 21 is intended to show a series of spots arranged side by side in twos in each cell, and of different colors. This, I believe, is impossible, for it is contrary to the law of bilaterality of color arrangement about the usual axis (HII, Figs. 6, 7).

In Fig. 22 there are several conditions which are impossible; e. g., an eye-spot situated upon a nervure is never seen in nature, also two spots originally side by side, as in cell III, *never rotate* around each other so as to come to lie one above the other. Spots often move, however, as shown by the arrows in cell IV, thus giving rise to fusions; or they may move away from each other, causing a wider gap between the rows. In cell Ib are shown two looped spots. One form (A) is quite usual, being found indeed in Cymothoe caenis Drury.[1] The other form of spot (D) is an impossibility, not being bilaterally symmetrical.

Fig. 23 illustrates other impossibilities in color-pattern, none of them, of course, being found in nature. For example, one never finds a row of slanting spots such as SS. Also one never sees a row of similar spots in *alternate* interspaces, such as is shown in DD, for this would be contrary to the law that similar spots are repeated in a row of adjacent interspaces. These last four diagrams

[1] See Cramer (1779–'82), Vol. 2, Plate 106.

(Figs. 20-23) have been introduced merely to give an idea of the curiously strict limitations which nature has imposed upon the differentiation of the color-pattern. Many beautiful effects might have been produced, such for example as that of alternate interspaces showing different colors, but this is not seen in nature.

It is interesting to recall the fact, that the colors themselves are impure and by no means so brilliant as they, perhaps, might have been, had Natural Selection been more severe in regard to color.

There is doubtless some physiological reason why spots almost invariably appear and disappear in the *middle of the interspaces*, and when we know more of the anatomical and histological conditions of the wing during the development of the colors, we may be able to discover it. It will be remembered that in the developing pupal wings of Callosamia promethea and Danais plexippus I found that the colors first made their appearance in the interspaces, and finally spread out so as to tinge over the nervures.

(4) *Origin of Color-Variations.* There is every reason to believe that all kinds of spots and bands, which are essentially only fused spots, may appear or disappear in any individual specimen without going through a long course of Natural Selection and slow phylogenetic differentiation. Darwin and Trimen ('71) and Bateson ('94) have demonstrated that this is true for eye-spots. In the Heliconidae I have found that bands and rows of spots are very variable in different specimens of the same species (see Plate **7**, Figs. 84-87).

There is a large and widely scattered literature recording the appearance and disappearance of colors and markings upon the wings of Lepidoptera. Limits of time and space prohibit my doing justice to it here, but it may be well to call attention to a very few of the more recent papers upon the subject. Many of the color-aberrations recorded in this list of papers may be due to the direct influence of environmental conditions upon the individual, but others are no doubt true sports or, to speak crudely, "congenital" variations, and might under favorable conditions of life become the ancestors of *new* varieties or species. It seems highly probable that new species often arise from just such sports in the manner so frequently and ably expounded by Bateson.

Partial Bibliography of Remarkable Color-Aberrations in
Lepidoptera.

Bairstow, S. D. '77. Ent. Mo. Mag., Vol. 14. p. 67. (Zygaena filipendulae.)
Bean, T. E. '95. Can. Ent., Vol. 27, p. 87–93, Plate 2. (Nemeophila petrosa
    and varieties.)
Benson, E. F. '83. Entomologist, Vol. 16, p. 210. (Arge galathea.)
Bleuse, L. '89. Feuille jeun. Natural., 19 Ann., p. 142–143.
Breignet, F. '90. Bull. Soc. Ent. France, (6), Tome 10, p. 29–30. (Thecla
    rubi; Melithaea athalia.)
Carrington, J. '78. Entomologist, Vol. 11, p. 97, Fig. (Cidaria suffumata.)
Carrington, J. '83. Entomologist, Vol. 16, p. 1, Fig. (Callimorpha dominula.)
Carrington, J. '88. Entomologist, Vol. 21, p. 73, Fig. Editorial Note. (Aretia
    caia.) And numerous other papers in the Entomologist.
Clark, J. A. '89. Entomologist, Vol. 22, p. 145–147, Plate 6. (Triphaena
    comes.)
Cockerell, T. D. A. '86. Entomologist, Vol. 19, p. 230–231. (Epinephele
    tithonus.)
Cockerell, T. D. A. '88. Entomologist, Vol. 21, p. 189. (Pieridae.)
Cockerell, T. D. A. '89. Entomologist, Vol. 22, p. 1–6, 13, 20–21, 26–29, 54–
    56, 98–100, 125–130, 147–149, 185–186, 243–245.
Dewitz, H. '85. Berlin. Ent. Zeitschr., Bd. 29, p. 142, Taf. 2. (Preeis amestris.)
Editors of Entomologist, '78. Entomologist, Vol. 11. p. 169–170, Plate 2.
    (Vanessa atalanta and several Lepidoptera.)
Edwards, W. H. '68. Butterflies of North America. (Numerous plates.)
Fettig, F. J. '89. Feuille jeun. Natural., 19 Ann., p. 81. (Variations of
    Lepidoptera in Alsace.)
Fitch, E. A. '78. Entomologist, Vol. 11, p. 50–61, Plate. (Colias edusa.)
Goss, H. '78. Entomologist, Vol. 11, p. 73–74, Fig. (Chelonia villica.)
Oberthür, C. '89. Bull. Soc. Ent. France, (6), Tome 9, p. 74–76.
Oberthür, C. '93. Feuille jeun. Natural., 24 Ann., p. 2–4.
Porjade, G. A. '91. Ann. Soc. Ent. France, (6), Tome 11, p. 597–598.
    Pl. 16. (Thais rumina.)
Richardson, N. M. '89. Ent. Mo. Mag., Vol. 25, p. 289–291. (Zygaena filipen-
    dulae.)
Scudder, S. D. '89. Butterflies of New England, p. 1243. (Bibliography of
    variations of Pieris rapae.)
South, R. '89. Entomologist, Vol. 22, p. 218–221, Plate 8. (Various Vanessidae.)
Speyer, A. '74. Stettiner Ent. Zeitung, Bd. 35, p. 98–103.
Thiele, H. '84. Berlin. Ent. Zeitschr., Bd. 28, p. 161–162, Fig. (Apatura iris.)
Tutt, J. W. '89. Entomologist, Vol. 22, p. 15, 160–161. (Melanic Agrotis
    corticea and pale variety of Lycaena bellargus.)

(5) *Climate and Melanism.* Lord Walsingham ('85), in his
presidential address before the Yorkshire Naturalists' Union, brought
forward the idea, that, although Arctic insects might be perfectly

able to withstand the most severe cold while in hibernation during the winter, it is of great importance for them to absorb as much heat as possible during the short summer. He placed several species of lepidopterous larvae upon a snow surface exposed to bright sunshine. The snow melted at different rates under the various larvae, and in two hours the darkest insect had sunk by far the deepest into the snow, proving that it was the best absorber of heat. This ingenious experiment of Lord Walsingham should be made the beginning of an extensive and careful research.

Chapman ('88) has shown that it may be of advantage to moths inhabiting wet regions to display dark colors, or become melanic. His observations were made upon Diamea flagella, and he says that upon one showery afternoon he observed that one side of the tree trunks was wet and dark in color; the other side being dry was paler. " As a consequence, the dark specimens of flagella were very conspicuous upon the dry portions, hardly visible on the wet, whilst with the ordinary form the conditions were reversed, those on the wet bark were conspicuous, those on the dry much less so." Perhaps the dull coloration of Arctic moths may be partially due to the effect of the somber background of rocks in the regions which they inhabit.

(6) *Relation between Climate and Colors of Papilios.* It is well known that the Lepidoptera in the Tropics display the richest variety and greatest number of colors. I have counted the colors exhibited by the 22 species of Papilio enumerated by Edwards as inhabiting North America north of Mexico, and also those which are displayed by the 200 species of Papilio named in Schatz's list as found in South America. The " colors" were determined by comparison with the colored plates in Ridgway ('86).

In this manner it was determined that the North American Papilios exhibit 17 colors, viz., black, brown, primrose-yellow, canary-yellow, sulphur-yellow, orange, white, greenish white, apple-green, cream-color, azure-blue, sage-green, rufous, pearl-gray, indigo-blue, iridescent blue, iridescent green.

On the other hand the South American Papilios exhibited 36 colors, viz., black, translucent black, brown, white, canary-yellow, citron-yellow, olive-yellow, primrose-yellow, chrome-yellow, straw-yellow, gamboge-yellow, cream-color, greenish white, apple-green, malachite-green, emerald-green, sage-green, slaty green iridescence, pea-green, azure-blue, iridescent Berlin-blue, indigo, pearl-blue, glaucous blue, salmon-buff, écru-drab, flesh-color, coral-red, rose-red, vermilion, rufous, geranium-red, geranium-pink, olive-buff, iridescent geranium-pink (as in P. zeuxis), and transparent areas.

As 200 species in South America display but 36 colors, while 22 in North America show 17, it follows that, while the number of species in South America is 9 times as great as in North America, the number of colors displayed is only a little more than twice as great. The richer display of colors in the Tropics, therefore, *may* be due simply to the far greater number of species, which gives a better opportunity for color-sports to arise, and not to any direct influence of the climate. The number of broods, also, which occur in a year is much greater in the Tropics than in the Temperate Zones, so that the Tropical species must possess a correspondingly greater opportunity to vary.

V. The Causes which have led to the Development and Preservation of the Scales of the Lepidoptera.

(1) *Experiments and Theory.* It is well known that the scales of Lepidoptera are morphologically identical with hairs. Indeed, a graded series from simple hairs, such as are found covering the body-surface of most Arthropods, up to perfectly developed flat scales bearing well differentiated striae may usually be found upon one and the same insect.

It is also remarkable that the color-bearing scales of beetles have been developed in the same manner as those of moths and butterflies, and that in this case also hairs have become differentiated into scales which are precisely similar in appearance to those of the Lepidoptera (see Dimmock, '83).

This is only another of the numerous instances met with in nature where similar conditions of selection have developed complex organs which are similar in appearance, though found in widely separated groups. A list of papers relating to the development of scales has been given by Dimmock ('83, p. 1-11).

Most of the hairs which cover the body-surface in Arthropods are true sensory structures, the axis of each of which is a protoplasmic process from a single cell of the hypodermis, which lies below the cuticula. They have probably been developed because the cuticula, being hard, chitinous, and inflexible, would serve but poorly as a tactile or sensory surface.

Of course no one would venture to ascribe any sensory function to the scales which cover the wing-membranes of the Lepidoptera. We may, however, make several more or less reasonable hypotheses concerning the probable uses of the scales, and by testing these suppositions arrive perhaps at some plausible explanation of their retention and the complex development which they have undergone.

(1) They may have caused the wings of the ancestors of the Lepidoptera to become more perfect as organs of flight, by causing the frictional resistance between the air and the wing-surface to become more nearly an optimum.

(2) The appearance and development of the scales may have served, as Kellogg ('94) has suggested, "to protect and to strengthen the wing-membranes."

(3) The present development of the scales may be due to the fact that they displayed colors which were in various ways advantageous to the insects.

Concerning the first of these three hypotheses, the wing has, broadly speaking, two chief functions to perform in flight. It must beat more or less downward against the air, and must, in addition, glide or cut through the air, supporting the insect in its flight. For the mere beating against the air a relatively *large* co-efficient of friction between the air and the wing might be advantageous; but for gliding and cutting through the air a *small* co-efficient of friction would certainly be an advantage. There must therefore be an optimum co-efficient of friction, which lies somewhere between these two.

In order to determine the co-efficient of friction between the wing and the air, use was made of a method which, in one form or another, has long been known to engineers; that is, of observing the ratio of damping of the vibrations of a pendulum.

It is well known that when a pendulum is swinging free, and uninfluenced by any frictional resistances, the law of its motion is expressed by the formula,

$$(I) \quad d = A \sin \frac{2\pi}{T} t$$

where d is the displacement of the pendulum from its middle position after the interval of time t, A is the maximum displacement and T the time of a complete vibration, back and forth. If, however, frictional resistances interfere, the formula becomes,

(2) $d = A e^{-Kt} \sin \frac{2\pi}{T_1} t$,

(3) Hence, $K = \dfrac{-\log d \; T_1}{A \log e \sin 2\pi \; t^2}$, or if $t = T_1$

(4) $K = \dfrac{-\log d}{A T_1 \log e}$

where K is a constant dependent upon the friction, e is the base of the Napierian system of logarithms and T_1 is the time of a complete vibration, which may be different from the T, representing the time of vibration when not under the influence of friction.

The plan was, then, to attach the wing of some large butterfly or moth to the end of a short, light pendulum in such a way that it would either fan against the air, or cut through it, and then to observe the ratio of damping of the pendulum's vibrations. A drawing of the pendulum with a wing attached is given in Plate 1, Fig. 3. The wing is here shown in the position for "cutting or gliding" through the air. It would be in the position for fanning against the air, if it were rotated 90°. The pendulum was made of brass and steel, the ends being of brass and the slender middle portion of steel. Its vibrations were read off upon an arc graduated in millimeters. The readings were certainly accurate down to 0.5 mm. The pendulum was hung upon a steel knife edge (N, N, Fig. 3), which rested upon firm level glass bearings. The pendulum was 24.21 cm. long, and weighed 19.61 grams. Its time of vibration (T_1) was 0.877 seconds. This rate of vibration was practically unaltered when a wing was fastened to the end of the pendulum, the reason being that the wings were very light, the heaviest, that of Samia cecropia, weighing only 0.038 grams. The wing to be experimented upon was fitted into a deep, narrow slot at the free end of the pendulum, and then cemented in by means of a little melted beeswax. It thus became a perfectly rigid part of the pendulum itself.

The pendulum with wing attached was deflected through a known arc, read off upon the millimeter scale, and its reading at the end of the first swing carefully observed. Then if A be the initial deflection, which we may call unity, and if d be the reading after the first swing, the ratio of damping is given by the expression $\dfrac{d}{A}$. In experimenting with a fore wing of Samia cecropia "fanning the air," it was found, as the mean of many trials, that this ratio of damping was 0.919, that is to say, the amplitude of the 2d swing was 0.919 as great as the amplitude of the 1st, that of the 3d only 0.919 as great as that of the 2d, and so on. The scales were then carefully removed from the wing-membranes, by means of a camel's hair brush, and by again testing the vibrations it was found that the new ratio of damping was 0.917. This is so near the value of the ratio of damping with the scales on (0.919), that it may be considered identical, the difference being due to errors of experimentation.

Hence we must conclude that the presence of the scales upon the wing-membrane has not altered, appreciably, the co-efficient of friction which would exist between scaleless wing-membranes and the air. The results indicate rather, that when the scales appeared upon the wings of the scaleless, clear-winged ancestors of the Lepidoptera, the co-efficient of friction remained unaltered. This tempts one to the further conclusions, that the co-efficient of friction between the air and the wings was already an optimum in these clear-winged ancestors before the appearance of the scales, and therefore that Natural Selection would operate to keep it unaltered.

A wing of Samia cecropia cut so as to give it the same shape and dimensions as one of Morpho menelaus, gave an identical damping ratio. I conclude that the co-efficient of friction may be the same for both moths and butterflies, at least for those which move their wings at about the same rate in flight.

It was found in the case of the Samia cecropia wing, that when it was vibrated in the position for " cutting through " the air, the ratio of damping was 0.991. It will be remembered that, when the wing " fanned " the air, this ratio was 0.917. We may find the ratio between the resistance encountered in " fanning " and that encountered in " gliding " through the air by substituting these values in equation (1), $\mathrm{K} = \dfrac{-\log d}{A T_1 \log e}$.

Thus for fanning, $\dfrac{d}{A} = 0.917$ and $T_1 = 0.877$. Making A unity,

$$\mathrm{K} = \dfrac{-\log 0.917}{0.877 \log e} = 0.1.$$

In cutting through the air, $\dfrac{d}{A} = 0.991$ and T_1 as before $= 0.877$.

Hence in this case $\mathrm{K} = \dfrac{-\log 0.991}{0.877 \log e} = 0.01.$

The wing, then, encounters at least 10 times the resistance in fanning that it does in gliding through the air. It should be said that this last experiment is somewhat crude, for the wing necessarily could not be made to cut the air with that delicate precision which is probably realized by the insect in flight. I should not be surprised, if in nature the insects encountered at least 20 times the resistance in beating the air, that they do in merely gliding through it.

Concerning Mr. Kellogg's supposition, that the scales were developed to "protect and to strengthen the wing-membranes," I will admit that they may serve in some slight degree to protect the wing-membranes from scratches, etc.; but I am unable to accept his conclusion, that they strengthen the wing-membranes, any more than that the shingles upon a roof serve to add strength to it. The wing-membranes themselves are tough, elastic, and not easily torn or scratched, and the scaleless wings of the Neuroptera and Hymenoptera are very rarely found torn or scratched in nature.

In 1858 Mr. Alexander Agassiz called attention ('59) to the fact, that "the nervures of the wings of butterflies are so arranged as to give the greatest lightness and strength; they are hollow, with their greatest diameter at the base of the wing, the point of greatest strain, their diameter gradually diminishing to the edge of the membrane. If a section be made across such a wing parallel to the axis of the body, we find very much the arrangement which has been experimentally proved by Fairbain and Stephenson as giving the greatest strength of beams, as exemplified in the tubular bridge. We find the strongest nervure placed either on or near the anterior edge of the upper wing; there is no such nervure on the lower wing, all being of nearly the same size, as such a one would have prevented the elasticity of the wing from assisting the flight to any considerable extent." Mr. Agassiz has informed me that he carried out an extensive series of experiments upon the rigidity of the wings of various species of Lepidoptera. He placed little platinum strips upon the wings and observed the extent of the bending produced. His results demonstrated that the Sphinx moths possess by far the strongest wings, and that the Danaoid and Acraeoid Heliconidae have very weak wings. The reason for this probably lies in the fact, that the Sphinx moths move their wings with great rapidity, while, according to Bates ('62) and all subsequent observers, the Heliconidae have a slow flight.

As the scales have been developed not because they aided the insects in flight or strengthened the wings, their retention must have been due to some other cause, probably to their displaying colors which were advantageous to their possessors in various ways. As Dimmock ('83) says, " it is only in insects where certain kinds of brilliant coloration have been developed that one finds scales." Indeed, I believe that the vast majority of the scales found in Lepidoptera are merely color-bearing organs. They probably first made their appearance upon small areas of the wings, perhaps adjacent to the body, and were merely colored hairs, similar to those of the surface of the body, which had grown out upon the wings. In this position they displayed some color which was of advantage to the insect; perhaps serving to render it less conspicuous than formerly. Under these circumstances they would naturally be preserved through the operation of selection until finally they became modified into true scales; just as the hairs in the Coleoptera have undergone a similar modification. If this be true, it is easy to see how they might spread out over the surfaces of the wings until the whole wing became covered with scales.

(2) *Summary of Conclusions.* The scales do not aid the insects in flight, for the wings have precisely the same efficiency as organs of flight when the scales are removed. The phylogenetic appearance and development of the scales upon the scaleless ancestors of the Lepidoptera did not in the least alter the efficiency of their wings as organs of flight. This efficiency of their wing surfaces was probably, therefore, already an optimum before the scales appeared. The scales do not appreciably strengthen the wing-membranes, that function being performed by the nervures. The majority of the scales are merely color-bearing organs, which have been developed under the influence of Natural Selection.

PART B.

COLOR-VARIATIONS IN THE HELICONIDAE.

I. General Causes which determine Coloration in the Heliconidae.

In 1861, after eleven years of study within the forests of South America, Bates read his, now classic, paper upon the life and habits of the Heliconidae of the Amazon region. In it he first brought forward his ingenious theory of Mimicry—a theory which, under the able interpretations of Wallace and Fritz Muller, and in more recent times, under the impetus of the zeal of their numerous disciples, has yielded so much that is of interest to scientific men.

The Heliconidae are, above all, creatures of the forest, and Bates found that the number of species increases as one travels inland from the Lower Amazons towards the eastern slopes of the Andes, so that the hot Andean valleys near Bogota, or in Ecuador, contain perhaps the greatest number. In their range they are restricted to the Tropics of the New World. Only two species, Dircenna klugii and Heliconius charitonius, extend so far north as the extreme Southern States of the United States, and none of them are found much further south than 30° S. Lat.

Bates and Felder first saw that the Heliconidae were naturally divided into two distinct groups. One, the Danaoid Heliconidae, consists of about twenty genera, all more or less closely related, and evidently an offshoot from the great universal family, the Danaidae, members of which are found in both Hemispheres. The other group, the true Heliconidae, is composed of two closely related genera, Heliconius and Eucides. They are allied in structure to the Acraeidae and hence their name, Acraeoid Heliconidae. Schatz and Röber ('85–'92, p. 105) say of the Acraeoid Heliconidae:—They are an offshoot of the great family Nymphalidae, which have undergone a remarkable development in the length of the fore wing, and in this respect have been developed in a direction parallel with the Danaoid Heliconidae. In their structure, however, they are quite distinct from the Danaoid group.

Schatz has proposed a new classification for the Heliconidae. He finds that the genera Lycorea and Ituna, which Bates included among the Danaoid Heliconidae, are very closely allied to the Danaidae, he therefore says that Lycorea should be placed among the Danaidae, while Ituna is clearly midway between the Danaidae and the Danaoid Heliconidae. Schatz proposes the name " Neotropidae " for the Danaoid Heliconidae. However, I think the name "Danaoid Heliconidae," being older and more descriptive of their relationship, should by all means be retained. In this paper I shall follow Bates's classification, and include among the Danaoid Heliconidae the twenty genera: Lycorea, Ituna, Athesis, Thyridia, Athyrtis, Olyras, Eutresis, Aprotopos, Dircenna, Callithomia, Epithomia, Ceratinia, Sais,

Scada, Mechanitis, Napeogenes, Ithomia, Aeria, Melinaea, and Tithorea. The Acraeoid Heliconidae will then consist of the two remaining genera, Heliconius and Eueides.

Staudinger ('84 '88) records 453 species belonging to the Danaoid group, and 150 belonging to the Acraeoid group.

Nearly all that we know concerning the early stages of the Heliconidae is due to Wilhelm Müller ('86). He gives figures and more or less complete descriptions of the early stages of Dircenna xantho, Ceratinia eupompe, Ithomia neglecta, Thyridia themisto, Mechanitis lysimnia, and also of Heliconius apseudes, H. eucrate, H. doris, Eueides isabella, E. aliphera, and E. pavana. Bates ('62, p. 596) says that he raised the larvae of Heliconius erato and Eueides lybia. Schatz and Röber ('85 '92) figure the larva and pupa of Ceratinia euryanassa. Edwards has given a detailed account of the early stages of H. charitonius.

Müller found that the larvae of the Danaoid group feed on various species of Solanum, while the genera Heliconius and Eueides feed upon the Passifloreae. The larvae are conspicuously colored, and often gregarious; they seem to take but little pains to hide themselves during the chrysalis stage, for Müller says that he has seen the silver-spotted, white chrysalids of Heliconius doris hanging in great numbers in the near neighborhood of the larval food plant. The mature insects also furnish a good example of what Wallace ('67) designated as "warning coloration," for their tawny orange and black wings are very conspicuous as they sail slowly around in circles, settling at frequent intervals in their lazy irregular flight.

Bates was the first to call attention to the circumstance that they often possess a rather strong and disagreeable odor, and in 1878 Fritz Müller confirmed this observation for a number of the Heliconidae. He found, for example, that the genera Ituna and Ilione have a pair of finger-like processes near the end of the abdomen, which can be protruded and then emit a rather disagreeable odor; and he also found that the Acraeoid Heliconidae, especially the females, possess a disgusting odor. Seitz ('89), however, examined about fifty species of Heliconidae and found that many of them appear to have no odor. For example, he says that Heliconius eucrate and Eueide dianasa have no odor, but that *some specimens* of Heliconius beskei, and Eueides aliphera have a horrid odor.

Whether they are odorous or not, it would seem that the Heliconidae have but few enemies to fear, for not one of the many skilled observers who have studied them in their native haunts has ever seen a bird attack them, and the only ground for believing that they are attacked rests upon the rather dubious evidence of a few specimens found by Fritz Müller having symmetrical pieces apparently bitten out of the hind wings. Belt ('74) observed that a pair of birds which were bringing large numbers of dragon-flies and butterflies to their young never brought any of the Heliconidae, although these were abundant in the neighborhood. In fact, Belt was able to discover only one enemy of these butterflies, and that was a yellow and black wasp, which caught them and stored them up in its nest to feed its young. The Heliconidae then, in spite of their weak structure, conspicuous colors, and slow flight, enjoy a peculiar immunity.

As is well known, Bates ('62) first called attention to the fact that the Heliconidae were " mimicked " or imitated both in color-pattern and shape of wings by a number of other genera of butterflies and even moths. Bates had no difficulty in showing that this mimicry might easily be explained upon the ground that the Heliconidae, on account of their bad taste and smell, were immune from the attacks of birds and other insectivorous animals, and that therefore it gave a peculiar advantage to a butterfly belonging to any other group not thus protected, to assume the shape and coloration of the Heliconidae ; for then the birds could not perceive any difference between it and the true Heliconidae. Bates found that fifteen species of Pieridae belonging to the genera Leptalis and Euterpe, four Papilios, seven Erycinidae, and among diurnal moths three Castnias and fourteen Bombycidae imitate each some distinct species of the Heliconidae occupying the same district. He also found that all of these insects were much rarer than the Heliconidae which they imitated. In some cases, indeed, he estimated the proportion to be less than one to a thousand. Wallace ('89, p. 265), who has added so much to our knowledge of this subject, aptly defines this kind of mimicry as an " exceptional form of protective resemblance."

But by far the most remarkable discovery made by Bates was the fact, that species belonging to different genera of the Heliconidae themselves mimic one another. Neither Bates nor Wallace was able to give any satisfactory explanation of the cause of this latter form of mimicry, for all of the genera of the Heliconidae are immune. They therefore supposed it to be due to " unknown local causes," or similarity of environment and conditions of life.

Thus the matter rested until 1879, when Fritz Müller brought out his well-known paper upon "Ituna and Thyridia, a remarkable example of mimicry," in which he showed that both of these genera are protected, yet they mimic each other. He also showed that this mimicry might be due to Natural Selection brought about in the following manner. It is possible that young birds, upon leaving the nest, are not furnished with an unalterable instinct which tells them exactly what they should and should not eat; so they may try experiments, and would then in all probability taste a few of the Heliconidae before finding out that they were unfit to eat. Müller then demonstrated that, if this supposition be true, it becomes a decided advantage to the various species of Heliconidae to resemble one another. His reasoning was as follows: Let it be supposed that the young and inexperienced birds of a region must destroy 1,200 specimens of any distasteful species of butterfly before it becomes recognized as such, and let us assume further that there are in existence 2,000 specimens of species A, and 10,000 of species B; then, if these species are *different* in appearance, each will lose 1,200 individuals, but if they resemble each other so closely that they cannot be distinguished apart, the loss will be divided pro rata between them, and A will lose 200, and B 1,000; therefore A saves 1,000 or 50% and B saves only 200 or 2% of the total number of individuals in the species; hence, while the relative numbers of the two species are as 1 to 5, the relative advantage derived from the resemblance is as 25 to 1.

Blackiston and Alexander ('84) have given a complete mathematical statement of Müller's law, and have come to the conclusion that, if the number of individuals destroyed is small compared with the number constituting the species, the relative advantage is inversely as the square of the original numbers; but if the number destroyed is large compared with the original number, the ratio of advantage is much greater than the inverse squares of the original numbers. Their deduction may be briefly stated as follows:—

Designation of Species	A	B
(1) Original number	$a >$	b
(2) Number lost without imitation	c —	$= c$
(3) Remainders without imitation	$(a-c)$	$(b-c)$
(4) Number lost with imitation	$\dfrac{a}{a+b}c$	$\dfrac{b}{a+b}c$
(5) Remainders with imitation	$a\left(1 - \dfrac{c}{a+b}\right)$	$b\left(1 - \dfrac{c}{a+b}\right)$
(6) Excess of remainders due to imitation, or "absolute advantage" (3)—(5)	$\dfrac{bc}{a+b}$	$\dfrac{ac}{a+b}$
(7) Ratio of excess to remainders without imitation (6): (3), =proportional advantage.	$\dfrac{c}{a+b} \times \dfrac{b}{a-c}$	$\dfrac{c}{a+b} \times \dfrac{a}{b-c}$
(8) Ratio of proportional advantage of B to proportional advantage of A.	\multicolumn{2}{c}{$= \dfrac{a\,(a-c)}{b\,(b-c)} = \dfrac{a^2}{b^2}\dfrac{\left(1 - \dfrac{c}{a}\right)}{\left(1 - \dfrac{c}{b}\right)}$}

It is evident, then, if c be small compared with a and b, that the proportional advantage of B is to the proportional advantage of A as a^2 is to b^2. If, however, the loss (c) is great compared with a or b, the relative gain for the weaker species becomes even greater than the ratio of the squares of b and a.

If it be true, then, that young birds, when they leave the nest, do *not* possess a directing instinct telling them what they should and should not eat, but actually do experiment to some extent upon various insects which they meet with, Müller's law is amply sufficient to account for the numerous cases of mimicry and remarkably close resemblances which are found among the species of the Heliconidae themselves.

Unfortunately no *direct* experiments have ever been made upon the feeding-habits of young South American birds, nor have the contents of their stomachs been examined. There have been a few experiments, however, which seem to support the idea that some animals do learn to associate an agreeable or disagreeable taste with the coloration and appearance of their prey. It is well known that Weismann ('82, p. 336–339) found that the black and yellow larvae of Euchelia jacobaeae were refused by the green lizard of Europe. He then introduced some young caterpillars of Lasiocampa rubi, which are very similar in appearance to those of Euchelia. The lizards first cautiously examined the larvae, and finally ate them. After this Weismann reintroduced the E. jacobaeae larvae and the lizards were seen to taste them, apparently mistaking them for the edible L. rubi caterpillars.

Poulton ('87) carried out a most careful and well-conducted research upon the protective value of color and markings in insects in reference to their vertebrate enemies. He experimented upon three species of lizards and a tree-frog. Poulton combines his results with those of other observers and presents them in the form of a table, which certainly supports the suggestion of Wallace ('67), that brilliant and conspicuous larvae would be refused as food by some at least of their enemies. Poulton also shows that a limit to the success of this method of defence (conspicuous larvae having unpleasant taste or smell) would result from the hunger which the success itself tends to produce. In the Tropics, indeed, where insectivorous birds and lizards are far more numerous than with us, and where competition for food is great among them, "we may feel sure that some at least would be sufficiently enterprising to make the best of unpleasant food, which has at least the advantage of being easily seen and caught." This last suggestion of Poulton certainly seems reasonable ; moreover, it has occurred to me that young birds, being but little skilled in the art of obtaining their food, might quite often be forced by hunger to try various kinds of insects, and perhaps even the Heliconidae themselves.

Beddard ('92, p. 153–167) reports the results of an extensive series of experiments carried out by Mr. Finn and himself upon marmosets, birds, lizards, and toads. He arrives at conclusions which are quite different from those of Poulton and others, but it appears to me that his experiments were by no means so critically performed as those of Poulton. He frequently threw larvae into a cage containing many birds and observed them struggle for the prey. It may well be, however, that a bird would be quite willing to swallow a very unsavory mouthful in order to prevent any of its companions from, apparently, enjoying it. However, Beddard found that toads will eat any insect without hesitation in spite of brilliant coloration, strong odors, or stings. He also found that birds and marmosets would often devour "conspicuously colored" larvae without any hesitation, and that some "protectively colored" or inconspicuous larvae were refused. There can be no doubt that many insectivorous animals pay but little attention to the colors of their prey ; for example, it is well known to anglers that trout and salmon will snap at the most gaudily colored "flies," which may or may not have any counterpart in nature.

The whole question of warning coloration will have to be made the subject of an extensive research upon both old and young insectivorous animals before we can safely arrive at any certain conclusions respecting it.

II. Methods Pursued in Studying the Color-Patterns of the Heliconidae.

No comparative study of the color-patterns displayed by the Heliconidae has ever been made. In fact, very few such studies have been carried out upon any Lepidoptera. The only works I know of are those of Eimer ('89) and Haase ('92) upon the coloration of the Papilios, and of Dixey ('90) upon the wing-markings of certain genera of the Nymphalidae and Pieridae. The family of the Heliconidae with its numerous species and comparatively simple coloration affords an excellent opportunity for such a research.

In making this study of the Heliconidae I was permitted through the kindness of Mr. Samuel Henshaw to make free use of the collection in the Museum of Comparative Zoölogy at Harvard. I also found the colored figures in the works of the following authors of great service : Hewitson ('56–'76), C. und R. Felder ('64–'67), Hübner ('06–'25), Humboldt et Bonpland ('33), Cramer (1779–'82), Staudinger ('84–'88), Godman and Salvin ('79–'86), and Ménétriés ('63) ; likewise the following shorter papers published in various serials : Bates ('63, '65), Butler ('65, '69, '69–'74, '77), Druce ('76), Godman and Salvin ('80), Hewitson ('54), Snellenen van Leeuwen ('87), Srnka ('84, '85), Staudinger ('82), and Weymer ('75, '84). I was thus enabled to examine the color-patterns of 400 (89%) of the species of the Danaoid group, and of 129 (86%) of the Acraeoid group, either from the insects themselves or from figures given by the authors named above. The remaining species were either inaccessible to me, or were so vaguely described as to be unavailable. A list of the species known to me is given in Table 28.

(1) *The Two Types of Coloration in the Danaoid Heliconidae.* It is very remarkable that the color-patterns of all of the Heliconidae

may be grouped into two very closely related types. To the one of these I have given the name "*Melinaea type*," for it is characteristic of most of the species of the genus Melinaea. It is well represented by Figs. 46, 48, 49, 51, and 55–57 (Plate **4**). The insects which belong to this type possess wings colored with *rufous*, *black*, and *yellow*.

The other type I designate as the "*Ithomia type*," for it is very characteristic of most of the species of the genus Ithomia. Figs. 47 and 52 (Plate **4**) afford examples of it. This type differs from the Melinaea in that the rufous and yellow areas upon the wings have become *transparent*.

There are, also, many species, found in numerous genera, which fall between these two types of coloration, for the yellow and rufous spots upon their wings have become translucent, so that one may speak of them as "translucent yellow" and "translucent rufous." These spots are, so to speak, in process of becoming transparent, but a few yellow or rufous scales still remain dusted over the otherwise clear spaces. Most of the Dircennas are good examples of this type (Fig. 54, Plate **4**).

Of the 400 species of the Danaoid Heliconidae, about 125 belong to the "Melinaea type." It is well represented by most of the species of the genera Lycorea, Athyrtis, Ceratinia, Mechanitis, and Melinaea. About 30 Ithomias and half a dozen Napeogenes also belong to it. About 160 species belong to the "Ithomia type," and of this number fully 120 belong to the genus Ithomia. The others are found in the genera Ceratinia, Napeogenes, Ituna, and Thyridia, and many of them resemble the Ithomias so closely that they are said to mimic them. About 100 species, some of which are found in almost all of the genera, are intermediate in their color-patterns between the Melinaea and the Ithomia types. The 15 remaining species are represented by Melinaea gazoria (Fig. 53, Plate **4**), Ceratinia eupompe, and a few Ithomias, such as Ithomia hemixantho. In these forms almost all color has disappeared, so that the whole wing has become of a uniform dull translucent yellow, bordered on the outer edges by a grayish black.

(2) *Detailed Description of the Melinaea Type of Coloration.* Figs. 46, 48, 49, 51, and 55–57 (Plate **4**) afford examples of this type of coloration. In these insects we find the proximal half of the central cell of the fore wing occupied by a rufous-colored area, which I call the "inner rufous." It is marked I in all of the figures,

Beyond the "inner rufous" we find a black spot, marked II in the figures. It usually occupies the middle region of the cell of the fore wing, and I have designated it as the "inner black." Beyond the "inner black," and occupying most of the outer portion of the cell of the fore wing, is a light-colored area, marked III in the figures. This area is rufous in color in Fig. 49, but it is usually yellow, as in Figs. 46, 48, 51, 54–57, and I have called it the "inner yellow." Beyond the "inner yellow," and occupying the extreme outer portion of the cell, lies the "middle black" (IV). In many species it is fused, as in Figs. 46–48, 56, 57, with the large black area, the "outer black" (VII), which occupies the greater portion of the outer half of the fore wing. Just outside of the cell beyond the "middle black" one finds a well-developed yellow area (V), the "middle yellow," and there is sometimes still another yellow patch beyond this, which is marked VI and called the "outer yellow." Finally, one often finds a row of white or yellow spots, the "marginal spots" (IX), lying very near the outer margin of the fore wing (see Figs. 47–49, 51, 54, 56). These spots are very well developed in the genera Ceratinia, Napeogenes, Ithomia, and Melinaea. One more very characteristic marking of the fore wing remains to be noticed; that is the longitudinal black stripe (VIII). It is also worthy of note that the front costal edge of the fore wing is almost always tinged with black.

The pattern of the hind wing is quite simple. The ground color is usually rufous and a "middle black" band (XI) runs across the middle of the wing. The outer edge is bordered by the "outer black" (XIII). Above the "middle black" band lies the "inner rufous" (X) of the hind wing, and below the "middle black" band one finds the "outer rufous" (XII) of the hind wing. One often finds a row of white or yellow dots within the outer black border of the hind wing, and these I designate as the "marginal spots" of the hind wing.

The *Ithomia type* of coloration, it will be remembered, may be derived from the Melinaea, by simply imagining the rufous and yellow areas to have become transparent. Also the outer black usually suffers a reduction so as to become only a rather narrow border along the outer margin of the fore wing. Thyridia psidii (Fig. 47) is a good example of this type. It will be seen that the black areas remain about the same as in the Melinaea type, but that the rufous and yellow have become transparent. The middle and outer yellow areas have also fused into a large transparent patch.

Ithomia sao (Fig. 52, Plate **4**) is another good example of the Ithomia type. In this particular species the "inner black" of the fore wing is absent, and the "middle black band" of the hind wing has disappeared. When we come to consider the other Ithomias, we shall find that in this genus it has probably fused with the marginal black of the hind wing.

I have made a record of the color-variations that affect the various characteristic areas just considered, and have recorded them for every one of the species of the Danaoid and Acraeoid Heliconidae known to me. As these records are too extensive for convenient inspection, I have condensed the results, and they will be found in Tables 1–27 inclusive. Thus, Table 1 gives the variations in color of the "inner rufous" area of the fore wing for each genus of the Danaoid Heliconidae; Table 2 records the variations of the "inner black"; Table 3 the "inner yellow" area, etc. In Table 1 we find, for example, opposite the genus Ituna, the number 2 in the column labeled "transparent." This indicates that in two species of Ituna the "inner rufous" area is transparent.

In order to facilitate the study of the color-patterns Dr. Davenport suggested that I make use of the ingenious projection method invented by Keeler ('93). This method consists in "squaring the wing" in the manner shown in Figs. 4 and 5 (Plate **1**). In Fig. 4 the large rectangle (A, B, C, D) just at the right of the figure of the *hind* wing represents a kind of Mercator's projection of the wing itself. The nervures 1^a, 1^b, 2, 3, etc., are represented by the vertical lines 1^a, 1^b, 2, 3, etc., on the rectangle A, B, C, D. In cells 1^a, 1^b, and 1^c (bounded by nervures 1^a, 1^b, and 2,) one finds a sinuous line winding across the middle of the cell. This line appears in the same relative position upon the rectangle A, B, C, D. The same is true of the eye-spot found in the cell bounded by nervures 2 and 3, and of all the other markings of the wings. The central cell of the wing itself is shown projected in the dotted rectangle E, F, G, H.

In the case of the *fore* wing (Fig. 5), the central cell of the wing is dotted, and is shown projected upon the similarly dotted area within the rectangle I, J, K, L. In other respects the method of projection is the same as in the case of the hind wing.

In this manner the colors displayed by various species of Danaoid and Acraeoid Heliconidae have been represented in color in Plates 5-8. Each large rectangle upon the left hand side of the Plate represents a hind wing, the small middle rectangles show the colors of the cell of the hind wing, and the right hand rectangles give the fore wings, all being projected in the manner illustrated in Figs. 4 and 5, Plate **1**. The chief advantage in Keeler's projection method lies in the fact, that similar areas in the projection of the wings lie vertically under or over one another, and thus by merely glancing up or down the plates one may observe the color-variations which occur in homologous cells of all the species represented.

III. General Discussion of the Color-Patterns and of Mimicry in the Genera Heliconius and Eueides.

Among the species of the genera Heliconius and Eueides we find remarkably little variation in venation, but great diversity in color-pattern of the wings, and in this respect they are very different from the Danaoid Heliconidae, where, it will be remembered, we find fully twenty different types of venation and only two types of color-pattern.

(1) *The Four Color Types in the Genus Heliconius.* Schatz and Röber ('85-'92) divide the species of the genus Heliconius into four groups based on color differences, as follows:—(1) the "Antiochus group" (Plate **4**, Fig. 50); (2) the "Erato group" (Fig. 60); (3) the "Melpomene group" (Fig. 59); and (4) the "Sylvanus group," a good example of which is Heliconius eucrate (Fig. 58, Plate **4**).

It will become apparent through an inspection of Figs. 50, 60, 59, and 58, which represent respectively, Heliconius antiochus, H. erato, H. melpomene, and H. eucrate, that the first three are quite closely related in color-pattern, while the fourth (H. eucrate) approaches very closely to the plan of coloration of the Melinaea type of the Danaoid Heliconidae. In fact this resemblance is so close that it may be safely said that the members of the "Sylvanus group," to which H. eucrate belongs, mimic the Danaoid Heliconidae.

The "Antiochus group" is represented by Heliconius antiochus (Plate **4**, Fig. 50, and Plate **5**, Fig 62). H. sara, H. galanthus, and H. charitonius (Plate **5**, Figs. 61, 63, 64) are also members of this group; other examples are H. apseudes, H. cydno, H. chiones, H. hahnesi, H. sappho, H. leuce, H. eleusinus, and H. clysonymus.

These species are characterized by their blue iridescence, and the narrow yellow or white bands upon the primaries; the hind wings are pointed at the outer apex, and the venation approaches the type found in Eueides aliphera. H. ricini (Plate 5, Fig. 66) is a good example of a form intermediate in coloration between group 1 and the "Erato group" (2).

The type of group 2 is Heliconius erato (Plate 4, Fig. 60, and Plate 5, Figs. 67 and 68). This group is closely allied to group 1 in its characteristics. A good connecting link between groups 1 and 3, the "Melpomene group," is H. phyllis (Fig. 65).

The third, or "Melpomene group," is represented by H. melpomene, H. callicopis, H. cybele, H. thelxiope, and H. vesta (Plate 6, Figs. 70–74, and Plate 4, Fig. 59). H. vulcanus, H. venus, H. chestertonii, H. burneyi, and H. pachinus are also examples of this group.

(2) *Mimicry between the Genus Heliconius and the Danaoid Group.* To Schatz's group 4, the "Sylvanus group," belong all those species of Heliconius which have departed widely from the coloration pattern of the other three groups, and have come to resemble various species of the genera Melinaea, Mechanitis, and Tithorea of the Danaoid Heliconidae. H. eucoma, H. eucrate, H. dryalus, and H. sylvana (Plate 8, Figs. 88, 89, 91, and 95) are good examples of group 4. By glancing at the diagrams on Plate 8 it will be seen that H. dryalus resembles Melinaea paraiya very closely; in fact, the likeness is so close that it is almost certain that no eye could distinguish between the two insects when they are upon the wing. Another startling resemblance is that between H. eucrate and Melinaea thera (Plate 8, Figs. 91 and 92); moreover, there is but little difference between the color-patterns of H. eucrate, Eueides dianasa, and Mechanitis polymnia (Figs. 91, 93, and 94). H. sylvana and Melinaea egina (Figs. 95 and 96) are also said to mimic each other. The resemblance certainly appears very close at a casual glance, yet when the colors are plotted, as in Figs. 95 and 96, the differences become quite apparent. H. claudia (Plate 5, Fig. 69) is a good connecting link between the Sylvanus group and the Melpomene group. In both the Melpomene and Sylvanus groups the venation has departed from the Eueides aliphera type, and the contour of the hind wings is much more rounded and elliptical than is the case in the Antiochus and Erato groups. (Compare Figs. 50 and 60 with Figs. 58 and 59, Plate 4.) There are rather less than twenty species which certainly

belong to the Sylvanus group; among them may be mentioned, in addition to those already spoken of, Heliconius numata, which resembles Melinaea mneme and Tithorea harmonia; H. zuleica, which resembles a Mechanitis and is a good copy of Melinaea hezia; and H. metalilis, which is said to mimic Melinaea lilis; there are also striking resemblances between

H. aurora and Melinaea lucifer; H. messene and Melinaea mesenina;
H. eucrate and Mechanitis lysimnia; H. hecalesia and Tithorea hecalesina;
H. hecuba and Tithorea bonplandii; H. ethra and Mechanitis nesaea;
H. formosus and Tithorea penthias; H. pardalinus and Melinaea pardalis;
H. telchina and Melinaea imitata; H. ismenius and Melinaea messatis.

Most remarkable of all perhaps is the close resemblance between Heliconius aristiona, Mechanitis methone, and Ithomia fallax of Staudinger. In fact, Staudinger states in his "Exotische Schmetterlinge" that he hesitated for some time to describe Ithomia fallax on account of its close resemblance to Hewitson's Mechanitis methone. Good lists of the Heliconidae which are said to mimic one another are given by Wallace ('89, p. 250, 251), and by Haase ('93*, p. 146, 147).

(3) *The Three Color-Types in the Genus Eucides.* In the genus Eucides we meet with three color-types represented by E. aliphera, E. thales, and E. cleobaea. These insects are distinctly smaller than the species of the genus Heliconius, and the yellow spots upon their primaries are more ocherous in color than in Heliconius. E. aliphera (Plate 6, Fig. 77) represents the most highly specialized color-type. Eucides mercaui (Fig. 76), however, is a good connecting link between the color-patterns of E. aliphera and E. thales (Fig. 75), and E. thales is almost identical in color-pattern with Heliconius vesta (Fig. 74).

The other type of Eucides is represented by E. cleobaea, E. dianasa, E. isabella, etc. (Plate 6, Fig. 78, and Plate 8, Fig. 93). These resemble the Sylvanus group of Heliconius or various Melinaeas and Mechanitis.

(4) *Detailed Discussion of Plates 5-8.* PLATE 5 is intended to illustrate the types of coloration found in the Antiochus and Erato groups of the genus Heliconius. In H. sara (Fig. 61) the wings are suffused with a dark blue iridescence, and some narrow yellow bands of color are found upon the primaries. In H. antiochus (Fig. 62) we find similar bands of color upon the primaries, but they are changed to white. H. antiochus may have descended

from an albinic sport of H. sara. In H. galanthus (Fig. 63)
the white areas have greatly increased in size, and the iridescent
blue has become much lighter. In H. charitonius (Fig. 64) we
find the wings crossed by yellow spots and bands, but in some specimens this yellow color exhibits a decidedly reddish tinge. The figure
of H. charitonius in Staudinger's "Exotische Schmetterlinge" illustrates this peculiarity; indeed, spots which are commonly yellow are
often found red, and *vice versa*. In H. phyllis (Fig. 65) we find
along the upper part of the diagram of the hind wing a yellow marking, and a similarly shaped red mark is found in its near ally, H.
thelxiope (Fig. 73, Plate **6**). The same is also true of H. ricini
(Fig. 66, Plate **5**).

H. erato (Figs. 67 and 68, Plate **5**, and Fig. 60, Plate **4**) is very
remarkable, for there are no less than four distinct color-types
exhibited by different individuals of this species; one of them (Fig.
67) shows the basal half of the hind wing marked by six red tongues
of color edged with iridescent blue, and there is a dark rufous
suffusion upon some parts of the fore wing. In other specimens
(Fig. 68) the red tongues of color which characterized the hind wing
of Fig. 67 are almost absent, and only the blue iridescence is left;
also there is no rufous to be seen upon the fore wing. In another
type the blue iridescence of the hind wing has become green, and in
still other specimens the yellow stripes upon the fore wing have
become white.

As one looks over the diagrams upon Plates **5**–**8**, it becomes evident that yellow frequently changes to white, for we often find one
or two species of a genus which exhibit white spots identical in shape
and position with spots which are yellow in most of the others. Good
examples of this are H. antiochus (Plate **5**, Fig. 62), Melinaea
parallelis and Ceratinia leucania (Plate **7**, Figs. 82 and 83); likewise
the white spot near the outer apex of the fore wing in H. eucrate
(Plate **8**, Fig. 91), which is yellow in many individuals. Yellow
areas are also frequently changed to rufous or red; thus the yellow
basal half of the hind wing of H. eucrate (Plate **8**, Fig. 91) is often
found of a rufous tinge in individual specimens of the species, and
among the specimens of this species in the Museum of Comparative
Zoölogy one can trace a gradation of this area from bright yellow
to rufous. H. claudia (Plate **5**, Fig. 69) is introduced in order to
exhibit some of the differences between the "Sylvanus" group, to
which it belongs, and the "Antiochus" and "Erato" groups.

PLATE 6 is intended to exhibit the characteristic color-patterns found in the Melpomene group and in the genus Eueides. Fig. 70 represents H. melpomene, and Fig. 71 its near ally, H. callycopis, in which the red area of the fore wing has become broken up, and some red spots have made their appearance near the base of the hind wing. In the next variety of H. melpomene, H. cybele (Fig. 72), it is remarkable that the pattern of the fore wing has come to resemble the Sylvanus type, and is identical in general plan of coloration with the fore wings of the Melinaeas or Mechanitis (see Figs. 84 or 85, Plate 7, or Figs. 92 or 94, Plate 8). In its close ally, H. thelxiope (Fig. 73), a still nearer approach to the Melinaea type has come about by the development of a black band across the middle of the hind wing, and one has only to imagine a general fusion of the seven club-shaped red stripes of the hind wing in Fig. 73, Plate 6, in order to produce exactly the Melinaea type as exhibited, for example, by Eueides eleobaea (Fig. 78). In this connection it is worthy of note that Bates ('62) showed that H. thelxiope was derived from H. melpomene, there being between the two many intermediate forms.

H. vesta (Fig. 74) is evidently a close relative of H. thelxiope, and what is still more worthy of note is, that it is almost identical in the general effect of its color-pattern with Eueides thales (Fig. 75)! The yellow spots upon the fore wing of E. thales are, however, duller in hue than are those of H. vesta, and the insects are somewhat different in size, H. vesta spreading 78 mm., while E. thales spreads only 66 mm. It will be noticed that the chief difference between the color-patterns of these two species lies in the fact, that, while the black stripes of the hind wings in H. vesta lie along the nervures, in Eueides thales they occupy the middle of the cells themselves. The general resemblance of the two color-patterns may of course be merely accidental. An easy explanation, however, is afforded by the theory of mimicry, for the two species look very much alike until one subjects their color-patterns to close analysis, when remarkable differences appear. E. thales (Fig. 75) may have been derived from some such form as E. mereani (Fig. 76), for one has merely to imagine a greater development of the black and a general deepening of the rufous upon the hind wing of E. mereani to make it resemble E. thales quite closely. Finally, in E. aliphera (Fig. 77) the black serrated border of the hind wing is still more reduced, and the black stripe which crosses the cell of the fore wing in E. mereani is not present.

PLATE **7** is intended to illustrate the peculiarities of color-pattern found among the Danaoid Heliconidae. Thyridia psidii (Fig. 79) is an example of the transparent type of color-pattern found among the Danaoid Heliconidae, and especially prevalent among the Ithomias. It will be seen by comparing Fig. 79 with the other figures upon Plates **7** and **8**, that the chief difference lies in the fact, that in this type both the rufous and yellow areas have become transparent. The black area of the fore wing has also suffered a reduction, especially along the outer margin of the wing. Incidentally it should be mentioned, that in this *particular* species the middle black band of the hind wing has become tilted up at a sharp angle, instead of crossing the wing horizontally. A life-size figure of the wings of Thyridia psidii is given on Plate **4**, Fig. 47.

In Napeogenes cyrianassa (Fig. 80) and Ceratinia vallonia (Fig. 81) portions of the usually yellow and rufous areas have become transparent.

The spots upon the fore wing of the Melinaeas are usually yellow, but in Melinaea parallelis (Fig. 82) they are white. It would seem that this form may have descended from some albinic sport. Ceratinia leucania (Fig. 83) resembles Melinaea parallelis so closely in general plan of coloration, that it is very difficult to distinguish between them, even when the two insects are seen side by side. Ceratinia leucania, however, is somewhat smaller than Melinaea parallelis. Both occupy the same region in Central America, and the specimens from which the diagrams were drawn came from Panama.

Figs. 84–87 are drawn from various specimens of Mechanitis isthmia, all from Panama. They are intended to give some idea of the range of individual variation which is met with in this extremely variable form. The contraction of the middle black band of the hind wing in this form has already been noticed in the general discussion of the laws of color-pattern (see page 184). In Fig 87 it will be seen that the inner yellow stripe which usually crosses the cell of the fore wing has become very narrow and changed to a rufous color. However, upon the under surface of the wing it still remains as a yellow stripe. Indeed, in most color-changes the upper side of the wing seems to take the initiative, the under surface being more conservative. This is not true, however, in the Ithomias, where the black areas of the under side of the wings often are found to be rufous in color, while they still remain of the normal black upon the upper surface. The colors of the under surface are, however, usually identical with those of the upper, though they are always *duller in hue*. This may be due to the fact, that the colors of the upper surface are more frequently seen than those of the lower, for these insects often float lazily along with their wings horizontally extended. The operation of Natural Selection would then be more severe with the upper surfaces than with the lower.

PLATE 8 gives an analysis of the color-patterns of some of the Heliconinae and those Melinaeas, etc., which they resemble. H. encoma (Fig. 88) is a good example of the Sylvanus type, and with its rufous, yellow, and black wings it is certainly a wonderfully close copy of the color-pattern found so commonly among the species of the genus Melinaea of the Danaoid Heliconidae.

Heliconius dryalus and Melinaea paraiya (Figs. 89, 90) resemble each other so closely in size, shape, and coloration, that it must be impossible to distinguish between them when the butterflies are in flight; yet an analysis of their color-patterns shows that there are considerable differences between them. The shape of the yellow bands upon the fore wings is quite different; the inner black spot within the cell is double in Melinaea paraiya, and there is also a row of white spots along the margin of the fore wing.

A much closer resemblance is found between H. encrate and Melinaea thera (Figs. 91 and 92), where the Heliconins is almost a true copy of the Melinaea.

The color-patterns of Eueides dianasa (Fig. 93) and Mechanitis polymnia (Fig. 94) are also very nearly the same. Both are common species in Brazil.

Heliconius sylvana is said by Bates and by Wallace to mimic Melinaea egina. It will be seen by reference to Figs. 95 and 96 that their color-patterns are quite different in detail, yet the insects look very much alike when placed side by side, and may easily be mistaken for each other when upon the wing. Melinaea egina is much more common than Heliconins sylvana.

IV. GENERAL DISCUSSION OF THE COLOR PATTERNS AND OF MIMICRY AMONG THE DANAOID HELICONIDAE.

(1) *The Origin of the Two Types of Coloration.* The character of the variation in the Danaoid Heliconidae is very different from that of the genera Heliconius and Eueides, for while there is great

diversity of color-pattern and very little variation in venation among
the species of the Acraeoid group, exactly the opposite condition is
met with in the Danaoid group, where we find at least twenty
different types of venation and only two types of color-pattern.
One of these types of coloration is well exemplified by most of
the Melinaeas (Fig. 48, Plate **4**), and I have therefore called it the
" *Melinaea* " type. The other type is exemplified by most of the
Ithomias (Figs. 47 and 52) and has been designated in this paper as
the " *Ithomia* " type. In the Melinaeas, it will be remembered, we
find the rufous and black wings crossed by bands of yellow ; while
in the Ithomias, on the other hand, the rufous and yellow areas have
become transparent, often leaving the wing as clear as glass, and the
black, which is so characteristic of the outer half of the wing in the
Melinaea type, has shrunk away until it has come to lie along the
outer margin of the wing only.

By a study of all the genera of Danaoid Heliconidae we gain light
upon the question of the origin of the " Melinaea" and " Ithomia "
types of coloration. As we have seen (page 198), the Danaoid
Heliconidae are an offshoot from the great family Danaidae. Indeed,
two of the genera, Lycorea and Ituna, are so closely related to the
Danaidae that Schatz and Röber ('85–'92) propose to include them
within that family. There can be but little doubt that Lycorea and
Ituna are remnants of the ancestral forms which long ago shot off from
the Danaidae to form the Danaoid Heliconidae ; and it is interesting to
note, that in these two patriarchal genera we find the two distinct
types of color-pattern which are exhibited by the Danaoid Heliconidae, for all of the five known species of Lycorea are good examples
of the Melinaea type (see Lycorea ceres, Fig. 46, Plate **4**), while the
four known species of Ituna all exhibit the transparent, or Ithomia,
type of coloration. In fact, in their color-patterns the species of
Ituna remind one of gigantic Ithomias. The species of Lycorea,
however, are colored very much after the pattern of the Danaidae,
and indeed they have departed but little from the type of the
members of the great family whence they sprang. On this account
I believe that the Melinaea type of coloration, which is so characteristic of the species of Lycorea, is phylogenetically older than
the Ithomia type.

In order to account for the origin of the Ithomia type, we may
assume that, shortly after the primeval forms of the Danaoid Heliconidae began to segregate out from the Danaidae, the species were

few and probably rare. Under these circumstances any given insect would gain but little advantage by resembling merely the general type of the coloration of its fellows. For the relative advantage gained by such imitation, according to Fritz Müller's law, increases inversely as the square of the fraction whose numerator is the actual number of the imitating form and whose denominator is the actual number of the imitated. Therefore when the insects were still rare there would be few to imitate and consequently but little advantage would be gained by the imitation. Imagine, for example, that a single insect happens to imitate the color-pattern of a group of 100, and that the advantage gained thereby is represented by the number 1; it is evident from Fritz Müller's law that, if it happened to imitate the coloration of a group of 1,000, its relative advantage would be 100 instead of 1. We see, then, that mimicry within the group of the Danaoid Heliconidae became an important factor only after the group was well established and the insects became common. During the early history of the race, then, there would be but little tendency towards conservatism of color-patterns, and when the "Ithomia" and "Melinaea" types of coloration made their appearance, they both survived and now serve as the patterns for mimicry; and this accounts very well for the remarkable fact, that there are no other types of coloration than these two to be found within the whole group with its 450 species!

(2) *Mimicry among the Danaoid Heliconidae.* The genus Ithomia with its 230 species is the dominant genus of the Danaoid group, and in nearly all of the other genera individual species are found which have departed widely from their generic type of coloration and have assumed the clear wings of the Ithomias. A good idea of how far these interesting individuals may depart from the coloration of their type may be gained by comparing Fig. 53, Plate 4, which represents Melinaea gazoria, with Fig. 48, which represents a typical Melinaea (M. paraiya). It is evident that Melinaea gazoria is startlingly like an Ithomia both in size and coloration, although it retains the venation and generic characteristics of a Melinaea.

In Mechanitis, which is the most independent genus of the Melinaea type of coloration, all of the species are fair examples of the Melinaea type, except Mechanitis ortygia Druce, from Peru. Druce ('76) in his description of this curious little species states in astonishment that it possesses the venation of a Mechanitis, but the size and coloration of an Ithomia!

It is quite remarkable that although the genera Melinaea and Mechanitis serve as models of mimicry for the Acraeoid Heliconidae, they should themselves mimic Ithomia.

The genus Ithomia is, however, the most independent of all the genera of the Danaoid group, and I know of remarkably few good instances in which an Ithomia has apparently departed from the coloration of its type to assume the guise of the Melinaeas. One good example of such a change, however, is afforded by Ithomia fallax of Southern Peru, which resembles either Mechanitis methone or Heliconius aristiona of Colombia (see page 210). There is apparently a difficulty in ascribing this resemblance to mimicry, for the imitator and imitated do not occupy the same geographical regions.

In direct contrast with the independence of the Ithomias stands the case of the genus Napeogenes; for Godman and Salvin ('79-'86) say of Napeogenes, that nearly every species mimics some Ithomia which occupies the same district; and thus almost the very existence of the genus would seem to depend upon its mimicry of Ithomia.

It is not the purpose of this paper to discuss, in detail, the numerous interesting cases of mimicry which are believed to exist between members of the Danaoid Heliconidae. An excellent discussion of such cases, and of the relationships of the various genera, has been given by Haase ('93ᵃ, p. 116-127).

V. Quantitative Determination of the Variations of the Characteristic Wing-Markings in the Acraeoid and Danaoid Heliconidae.

(1) *Variations of "Inner Rufous" Areas of the Fore and Hind Wings.* Table 1 gives the color-variations which are exhibited by the "inner rufous" area of the fore wings in the Danaoid Heliconidae. This area is marked I in all of the figures upon Plate 4. We learn from an inspection of Table 1 that this area is rufous in color in 124 species of the Danaoid Heliconidae, transparent in 152, black in 24, and that in the remainder it is more or less translucent, and of either a yellowish or rufous tinge.

Table 10 shows the variations which come over the "inner rufous" area of the hind wings of the Danaoid Heliconidae. This area is marked X in the figures upon Plate 4. It is apparent at a glance that the variations which affect the inner rufous areas of both fore and hind wings are very similar. In order to exhibit this fact graphically, the color-variations have been laid off upon the diagram, Fig. 97, Plate **9**. The base line is marked at equal intervals with the words "rufous," "translucent rufous," "translucent, slightly rufous," "transparent," etc., and the ordinates show the number of species which exhibit the various colors, rufous, translucent rufous, etc. For example, at the point "translucent rufous" we find that the ordinate is 23; this indicates that in 23 species the area is translucent rufous in color. The points thus found upon the ordinates are successively joined by straight lines forming a zig-zag figure. The full line represents the fore wing, and the dotted line the hind wing, and it becomes clearly evident from the closeness of these two zig-zag lines that the color of the inner rufous area of the fore wing (area I, Plate **4**) is almost always sure to be identical with that of the inner rufous area of the hind wing (area X, Plate **4**). We see, therefore, that whatever color-variation affects the inner rufous area of the fore wing, this area in the hind wing is almost always affected in the same manner.

Fig. 99, Plate **9**, is derived from Tables 15 and 24, which show the color-variations in the fore and hind wings of the genera Heliconius and Eueides. It is seen that here also the colors of these two areas in both the fore and hind wings are almost always identical. We here meet with one of those interesitng physiological laws which are independent of Natural Selection, and the meaning of which remains a mystery, for surely we can see no reason on the ground of adaptation why similar areas upon both fore and hind wing should bear similar colors.

(2) *The "Inner Black" Spot.* Table 2 shows the presence or absence of the "inner black" spot in the Danaoid Heliconidae. This spot is marked II in the figures upon Plate **4**. When present, it is always black in color and is usually found occupying the middle region of the cell of the fore wing. The table shows that it is about an even chance whether it be present or not, for it is absent in 210 species and present in 190. In the genus Ithomia, however, it is present in only one third of the species. What is most worthy of note concerning it, is the fact that it almost always appears, when present, as a single spot. Indeed, it appears as a double spot in only 7 species, and 5 of these belong to the genus Melinaea. A good example of its appearance as a double spot is found in Melinaea paraiya (Fig. 48, Plate **4**). It will be remem-

bered that there are 450 species in the Danaoid group; 25 of
these belong to the genus Melinaea; yet among these 25 we find
5 exhibiting this marking as a double spot. Assuming that the
doubling of this spot has arisen in each species as a sport, and that
such a sport is as likely to appear in one species as in any other of
the Danaoid group, then the chances against five such sports
appearing among the 25 Melinaeas is $\frac{450\times449\times448\times447\times446}{25\times24\times23\times22\times21}$, or about
2,830,000 to 1. Indeed, it is probable that all five of the species
of Melinaea which exhibit the doubling of this spot are descendants of a *single* ancestor in which it appeared for the first time
double, for the mathematical chance that one such ancestor should
appear among the Melinaeas, rather than in any other genus, is
evidently 1 in $\frac{450}{25}$, or one chance in eighteen. The chance against
two such unrelated ancestors is, however, $\frac{450\times449}{25\times24}$, or about 336 to
1, and the chance against three is $\frac{450\times449\times448}{25\times24\times23}$, or 6,560 to 1, etc.

By reference to Table 16 we find that in the genera Heliconius
and Eueides the inner black area is black or iridescent blue in all
of the species of Heliconius, but absent in 5 of the 18 species of
Eueides known to me. These 5 include Eueides aliphera and its
allies. Now there are 150 known species of the Acraeoid Heliconidae, and 24 of these belong to the genus Eueides; so it is evident
that the mathematical chance against the supposition that five sports
arose independently in the genus Eueides, in which the inner black
was absent, is given by $\frac{150\times149\times148\times147\times146}{24\times23\times22\times21\times20}$, or 13,900 to 1. It is therefore probable that the five Eueides lacking the inner black are
the descendants of a single ancestor.

(3) *Variations of the "Inner Yellow" and "Middle Yellow"
Areas.* Tables 3 and 5, and diagram Fig. 98, Plate **9**, show the
color-variations of the "inner yellow" and "middle yellow" areas
in the fore wings of the Danaoid Heliconidae. These areas are
marked III and V, respectively, in the figures upon Plate **4**. The
"inner yellow" area, it will be remembered, occupies the outer portion of the cell of the fore wing; while the "middle yellow" is found
in the region just beyond the outer limits of the cell. The two areas
are often fused together as in Figs. 47, 48, 50, 51, and 55, Plate **4**.
The inner yellow area is usually smaller than the middle yellow,
and a comparison of Tables 3 and 5 will show that it is much more
frequently obliterated by the encroachment of the rufous or black

areas which surround it; for example, while the middle yellow is rufous in color in only 14 species, the inner yellow is rufous in 56; also the inner yellow area, being usually smaller and less conspicuous than the middle yellow, is less important in cases of mimicry, and the diagram Fig. 98, Plate **9**, shows that it is much more variable in color than the middle yellow. The full zig-zag line in this figure represents the color-variations of the inner yellow, while the dotted zig-zag line gives the color-variations of the middle yellow. As there are nine color-variations displayed by each of these two areas, and as there are 400 species of the Danaoid Heliconidae recorded by me, it becomes evident that, if there were no color preferences displayed by these areas, there would probably be about $\frac{400}{9}$, or 44.4, species which would display it as rufous, 44.4 translucent, 44.4 yellow, etc. The heavy, straight, dotted line (Fig. 98, Plate **9**) represents this ideal condition, which would be approximately realized were one color as likely to occur as another in the respective areas. Now it is evident from an inspection of the figure, that the full zig-zag line, which represents the color-variations of the "inner yellow," approaches the straight line condition more nearly than does the dotted zig-zag line, which represents the middle yellow.[1] The inner yellow is therefore more liable to color-variations than the middle yellow; and this is what we should expect on account of its comparatively small size and its consequent inconspicuousness as a characteristic marking in cases of mimicry.

A comparison of Figs. 97 and 98, Plate **9**, is interesting, for it shows that the color-variations of the inner rufous are quite similar to those of the inner yellow and middle yellow. This serves to illustrate the close physiological relationship which exists between rufous and yellow. The two pigments are probably closely related chemically, for every ordinarily rufous area is sometimes found to be yellow, and vice versa. Yellow areas also often change to white. Rufous, yellow, and white are evidently closely related color-variations.[2]

[1] This is not true for one color, white.

[2] It may be well to mention here that the black areas upon the wings are subject to very little color-variation. In some cases, however, especially upon the under surfaces of the wings in Ithomia, the black has changed to a rufous or russet color. For example, Table 4 shows that the middle black area (IV in the figures upon Plate 4) is rufous in only 12 species out of the 400 which are recorded, and all of these 12 are Ithomias. Also Tables 7 and 13 show that the outer black of the fore wing, and the outer black of the hind wing are russet in 22 and 11 species, respectively. Evidently, black is a far more conservative color than rufous, yellow, or white. Probably black is also quite different from the other pigments chemically.

Tables 17 and 19 show the color-variations affecting the "inner yellow" and "middle yellow" areas of the fore wing in Heliconius and Eueides. There is but little difference between the two tables, except that in 15 species of Heliconius the inner yellow is suffused with black or blue, while the middle yellow is never suffused by the outer black which surrounds it. Fig. 100, Plate **10**, exhibits graphically the color-variation of these two areas. The "inner yellow" is represented as a full line, and the "middle yellow" as a dotted zig-zag. It is evident that here also the inner yellow is more variable in color than the middle yellow, for not only does the inner yellow area display two more colors, but its chart is a flatter zig-zag.

(4) *Variations of the "Middle Black" Mark of the Fore Wing.* Table 4 shows the color-variation of the middle black mark (area IV in figures upon Plate **4**). This marking lies along the extreme outer border of the central cell of the fore wing. It is small in area, but is rendered very conspicuous from the fact that it is situated between the inner yellow and middle yellow markings. In spite of its small size, however, it is a remarkably permanent marking, for Table 4 shows that it is absent in only 20 out of 400 Danaoid Heliconidae. In these 20 it has been obliterated by the fusion of the inner and middle yellow areas. It is worthy of note that in 12 Ithomias it has become rufous in color. This change to rufous is the only color-change which the black areas of the wings ever display.

Table 18 shows the variations of the middle black area for Heliconius and Eueides.

(5) *Variations of the "Outer Yellow" Area of the Fore Wing.* Table 6 shows the variations which affect the outer yellow area of the fore wings in the Danaoid Heliconidae. This area is marked VI in the figures upon Plate **4**; it lies beyond the region of the middle yellow, but is usually more or less fused with it. Table 6 is only approximately correct, owing to the difficulty in many cases of deciding whether the middle and outer yellow be really fused or not. It will be seen that in the genus Ithomia the middle and outer yellows are wholly fused in about 200 species. This is one of the marked characteristics of this very independent genus.

Table 20 shows the color-variations of the outer yellow area in Heliconius and Eueides. This marking is present in 81 and absent in 48 of the Acracoid group. It is much more widely separated from the middle yellow than is the case in the Danaoid group.

(6) *The relative Permanency of the Black Areas upon the Fore and Hind Wings.* A study of the relative permanency of the various characteristic black markings upon the wings is of interest, for, if the generally accepted idea concerning the prevalence of mimicry within the group of the Danaoid Heliconidae be true, we should expect the most conspicuous markings to be the most permanent, for they are evidently of the most importance for mimicry. This is, however, not the case for the black markings. A good example of this fact is afforded by a comparison of the relative permanency of the black streak which extends along the extreme costal edge of the fore wing with the inner black spot (II in figures on Plate **4**). The inner black spot is certainly a more conspicuous marking than this narrow black streak along the costal edge; yet it is much more variable, for Table 2 shows that it is present in 210 and absent in 190 of the 400 Danaoid Heliconidae. In other words, it is about as likely to be present as absent. The black streak upon the costal edge, on the other hand, is much more permanent, for it is absent in only 52 species out of the 400.

Another good example of the inaccuracy of the supposition that large and conspicuously colored areas are always less variable than small ones, is derived from a comparison of the relative variability of the large outer black of the fore wing with the small outer black of the hind wing. Although the outer black area of the fore wing is usually much larger and more conspicuous than the outer black margin of the hind wing, it is more variable in color, for it is rufous in 22 species, while the outer black of the hind wing is rufous in only 11, out of the 400.

In general, however, large colored areas are more permanent than small ones, as was found in the case of the inner and middle yellow areas (see page 220). Indeed, a good instance of this greater variability of small color areas is afforded by the longitudinal black stripe marked VIII in the figures of Plate **4**, for this is more variable than the larger outer black area of the fore wing.

(7) *The "Middle Black Stripe" of the Hind Wing.* In the genus Ithomia the middle black stripe (XI, Plate **4**) has migrated downward, so that in many species it has become fused with the outer black margin, as in Ithomia sao (Fig. 52, Plate **4**). In other cases there is still to be seen a narrow line of rufous color between the middle black band and the outer black margin of the hind wing. Such is the case in Ithomia nise (Fig. 54, Plate **4**). In

many other cases the outer black and middle black are completely fused, so far as the upper surface of the wings is concerned; but, if one examines the under surface of the hind wings, it will be found that a narrow rufous streak still persists between the middle black band and the outer black margin of the hind wing.

(8) *Variations of the Marginal Spots of the Fore Wing.* The marginal spots are found very near the outer margin of the fore wing; they are usually either yellow or white, but in some few cases they are rufous. It appears from Table 9 that they are present in 146 and absent in 254 species of the 400 Danaoid Heliconidae known to me. Fig. 101, Plate **10**, shows graphically the manner in which these spots occur in those species which possess them. It is evident from this curve that the number of these spots is not determined merely by chance, for they show a marked tendency to appear either as 2 or 3, or as 6 or 7 spots. It is due to this fact, that there are two maximum points upon the diagram Fig. 101, Plate **10**. In those species which exhibit the "2- or 3-spot" condition, the spots are found near the front apex of the fore wing. In the "6- or 7-spot" condition they lie all along the outer margin of the fore wing, one spot in each cell. In the genera Ithomia, Napeogenes, and especially in Ceratinia these marginal spots have become large and conspicuous ornaments. (See Fig. 49, Plate **4**.)

Table 22 shows the manner of appearance of these spots in the genera Heliconius and Eueides. They are found in only 26 species of the 129 known to me; and this number is far too small to warrant general conclusions concerning the order of their appearance.

(9) *The Marginal Spots of the Hind Wing.* Table 14 illustrates the manner in which the marginal spots of the hind wings make their appearance. They are absent in 279 and present in 121 of the 400 species of the Danaoid group. Thus they occur rather less frequently than the marginal spots of the fore wing. In the 121 species in which these spots are found they show a decided tendency to appear either as 4 or as 5 spots. Fig. 102, Plate **10**, is a graphic representation of the distribution of these spots, derived from Table 14. It appears that the outline of the figure approaches a probability curve, and is approximately symmetrical about the mean ordinate (A, B), situated at 4.54.

VI. Comparison of the Color-Variations of the Papilios of South America with those of the Heliconidae.

In order to emphasize the peculiarities of the coloration of the Heliconidae, I will conclude by instituting a comparison between their variations and those of the South American Papilios. There are about 200 species of Papilio in South America, and these display in all 36 distinct colors. The colors have been determined by reference to the plates in Ridgway's "Nomenclature of color for naturalists." A list of the colors which are displayed by these Papilios has already been given upon page 191.

By exercising a very fine discrimination in distinguishing color we may count 15 distinct colors which are displayed by the 450 members of the Danaoid Heliconidae, as follows: black, brown, translucent black, sulphur-yellow, canary-yellow, citron-yellow, primrose-yellow, yellow-rufous, reddish rufous, rufous, white, translucent yellow, translucent rufous, transparent areas upon the wings, transparent areas which display iridescence. We see, then, that while the 200 species of Papilio display 36 different colors, the 450 Danaoid Heliconidae exhibit only 15. In other words, the *numbers* of the *species* and of the *colors* are almost in inverse ratio in the two groups; for while the Papilios are only $\frac{4}{9}$ as numerous as the Danaoid Heliconidae, they display almost $\frac{1}{4}$ times as many colors; and this is all the more remarkable when we remember that the general class of coloration in the Papilios and Danaoid Heliconidae is apparently the same. That is to say, in both groups we find all of the species displaying decidedly conspicuous colors, the coloration of the upper surfaces of the wings being in both rather more brilliant than that of the lower surfaces, but without essential differences in color-pattern. Nor is there an attempt in either case at protective resemblances, such as the imitation of the coloration of bark, leaves, etc. The color-patterns of the Papilios are, moreover, extremely complex, and upon comparing the different species, there are seen to be frequent fusions and obliterations of the characteristic markings, so that Haase ('93), who has made an extensive study of their color-patterns, is forced to divide them into many small groups of a few species each. The variation in the form of the wings is also very great among the Papilios, for while P. protesilaus possesses upon its hind wings, long tail-like appendages, the hind wings of P. hahneli are rounded off and without marked appendages.

There is, apparently, but one important respect in which the Danaoid Heliconidae are more variable than the Papilios, and that is size. For example, Lycorea ceres, which is probably the largest of the Danaoid group, has 2.2 times the spread of wing of Ithomia nise, which is one of the smallest (see Plate **4**, Figs. 46 and 54). The largest Papilio, P. androgea, on the other hand, spreads only 2.16 times as much as the smallest, P. triopas.

There is another minor respect in which the color-patterns of the Papilios are different from those of the Heliconidae. In the Heliconidae the fore wing slightly overlaps the hind wing, and that portion of the hind wing which is hidden from view is always dull in color (see Plates **5-8**). In the Papilios, however, the fore wing does not overlap the hind wing to such an extent as in the Heliconidae, and it is worthy of note that the costal edges of the hind wings in the Papilios are as brilliantly colored as are any other portions of the wings.

It is difficult to account for the remarkable conservatism in respect to color-variations among the Heliconidae, unless we resort to the explanation afforded by the theory of mimicry; for, while there is such remarkable simplicity and uniformity of color-pattern throughout the whole group of the Heliconidae, *individual variations* are very common. In the collection at the Museum of Comparative Zoölogy, for example, one finds a regularly graded series of specimens of Heliconius cuerate; at one end of this series the "inner rufous" area of the hind wing is bright yellow, and at the other end it is rufous; intermediate specimens are found in which this area is yellow, but dusted over with rufous scales. Also the "middle black band" of the hind wings in Melinaea parallelis is very variable, some specimens showing it broken in the middle (Plate **7**, Fig. 82), and others having it as an entire band. I have also seen one specimen of H. burneyi in which the commonly yellow spots upon the under surface of the wings were changed to white. Another good instance of individual variability is afforded by H. phyllis (Plate **5**, Fig. 65), for in this species the series of small red spots sometimes found just below the yellow band upon the hind wing is very variable, and more often absent than present. Still other instances of individual variability are seen in the yellow stripes upon the wings of H. charitonius (Plate **5**, Fig. 64), which are often found tinged with rufous. Also the remarkable diversity in Mechanitis polymnia, and M. isthmia (Plate **7**, Figs. 84-87) are other examples which show that there is no lack of individual variability among the Heliconidae. Yet the Danaoid species as a whole vary but little from the two great types of coloration represented by Ithomia and Melinaea, and in this respect they are very different from the Papilios, where we find a great many color-types and great diversity in shape of wings. Surely there must be some cause for the remarkable fact that the Danaoid Heliconidae with their 453 species should display but two types of color-pattern. I can think of but one explanation, which is afforded by Fritz Müller's theory of mimicry.

In conclusion it gives me pleasure to thank those friends whose generous aid and kindness have done so much to render the prosecution of this research a pleasure to me. I wish to express my gratitude to Dr. Charles B. Davenport, who is the real instigator and promoter of this research; to Mr. Samuel Henshaw, to whom I am indebted for numerous kindnesses, and who placed at my disposal the extensive entomological collections and library of the Museum of Comparative Zoölogy at Harvard; to Prof. Edward L. Mark for his kindness in revising the manuscript of this paper, and for the numerous valuable suggestions which he has made; to Dr. Samuel H. Scudder, to whom I am grateful for much kind advice and for the use of rare works in his library; to Prof. Ogden N. Rood for his valuable suggestions in regard to the spectroscopic apparatus; to Dr. Alpheus Hyatt for his valued and kind advice, and to my father, Prof. Alfred M. Mayer, for the use of Maxwell's discs and the direct-vision spectroscope.

PART C.

GENERAL SUMMARY OF RESULTS BELIEVED TO BE NEW TO SCIENCE.

(1) The great majority of the colors of Lepidoptera contain a surprisingly large percentage of black (p. 172).

(2) The colors displayed by the scales are not simple, but compounded of several different colors (p. 173).

(3) The pigments of the scales of Lepidoptera are derived by various chemical processes from the blood, or haemolymph, of the

pupa. The pupal blood of the Saturnidae is a proteid substance containing egg albumen, globulin, fibrin, xanthophyll, orthophosphoric acid, iron, potassium, and sodium (p. 176).

(4) In Callosamia promethea and Danais plexippus the pupal wings are at first perfectly transparent, then white, then impure yellow, excepting upon those portions which are destined to remain white in the mature wing. The mature colors then begin to appear near the central areas of the wings and *between* the nervures. Last of all, the nervures themselves become tinged with the mature colors. The central portions of the wings acquire their mature colors before the outer and costal edges, or the root of the wing adjacent to the body (p. 178, Plate **3**).

(5) The white stage in the development of color in the pupal wings represents the condition in which the scales are perfectly formed but lack the pigment which is destined to be introduced later (p. 178). (See, also, Mayer, '96, p. 230.)

(6) Dull ocher-yellows and drabs are, phylogenetically speaking, the oldest pigmental colors in the Lepidoptera. The more brilliant colors, such as bright yellows, reds, and pigmental greens, are derived by complex chemical processes and are, phylogenetically speaking, of recent appearance (p. 178). (See, also, Mayer, '96, p. 232.)

(7) While the number of species of Papilio in South America is 9 times as great as in North America, the number of colors which they display is only twice as great. Hence the greater number of colors displayed by the tropical forms may be due simply to the far greater number of the species, and not to any direct influence of the climate (p. 191).

(8) The following laws control the color-patterns of butterflies and moths: (a) Any spot found upon the wing of a butterfly or moth tends to be bilaterally symmetrical, both as regards form and color; and the axis of symmetry is a line passing through the center of the interspace in which the spot is found, parallel to the longitudinal nervures (p. 183). (b) Spots tend to appear not in one interspace only, but in homologous places in a row of adjacent interspaces (p. 183). (c) Bands of color are often made by the fusion of a row of adjacent spots, and, conversely, chains of spots are often formed by the breaking up of bands (p. 183). (d) When in process of disappearance, bands of color usually shrink away at one end (p. 184). (e) The ends of a series of spots are more

variable than the middle. This is only a special case of Bateson's ('94) law (p. 185). (f) The position of spots situated near the outer edges of the wing is largely controlled by the wing-folds or creases (p. 185).

(9) The scales in Lepidoptera do not strengthen the wings or aid the insects in flight. The vast majority of the scales are merely color-bearing organs, which have been developed under the influence of Natural Selection. The phylogenetic appearance and development of scales upon the originally scaleless ancestors of the Lepidoptera did not alter the *efficiency* of their wings as *organs of flight*. It is probable, therefore, that this efficiency was an optimum before the scales appeared (p. 197).

(10) A systematic study of the Danaoid Heliconidae demonstrates that their color-patterns can be placed in two types. Type 1, the more complex, is closely related to the coloration of the Danaidae from which the Danaoid Heliconidae sprang, and is therefore, phylogenetically speaking, the older type of coloration. This type is characteristic of the genera Lycorea, Melinaea, and Mechanitis, and I have called it the "Melinaea" type. It is characterized by the fact that the wings are rufous and black in color, and crossed by a definite system of yellow bands. Type 2, the "Ithomia" type, is characteristic of the genera Ithomia, Ituna, and Thyridia. The "Ithomia" type has been derived from the "Melinaea" by the originally rufous and yellow areas upon the wings having become transparent (p. 204).

(11) The phylogenetic origin of the "Melinaea" and "Ithomia" types of coloration can be accounted for upon the supposition, that when the species of the Danaoid Heliconidae began to segregate out from the Danaidae they were for a time rare (p. 215). ·

(12) A record of the characteristic markings upon the wings of the Danaoid and Acraeoid Heliconidae shows that, physiologically speaking, the colors red, rufous, yellow, and white are closely related, and that black is quite distinct from these, being the least variable color of all (p. 220).

(13) In both the Danaoid and Acraeoid Heliconidae, whatever color-variation affects that part of the *fore* wing which is adjacent to the body of the insect, almost always the same color-variation affects the homologous area of the *hind* wing in a similar manner (p. 218, and Fig. 99).

(14) The smaller yellow spots upon the wings of the Heliconi-

dae are more liable to color-variations than are the larger ones. This is what we should expect, if the theory of mimicry be true; for large spots are more conspicuous, and therefore their preservation is more important (p. 220). This rule, however, does not hold for the black markings of the wing (p. 222).

(15) The mathematical chance against five similar and independent color-sports arising in the genus Melinaea is as 2,830,000 to 1. Hence, the five Melinaeas which display the "inner black" as a double spot are probably descended from a single ancestor (p. 219).

(16) The marginal spots of the fore wing in the Danaoid Heliconidae show a marked tendency to appear either as 2 or 3, or else as 6 or 7 spots (p. 223, Fig. 101). The marginal spots of the *hind* wing show a marked tendency to appear either as 4 or 5 spots (p. 223, and Fig. 102).

(17) The 200 species of Papilio in South America display 36 distinct colors, while the 450 species of Danaoid Heliconidae exhibit only 15. Hence the numbers of the *species* and of the *colors* are almost in inverse ratio in the two groups. This may be explained by the fact, that the Danaoid Heliconidae mimic one another, while the Papilios do not (p. 224).

(18) The colors are dull upon those portions of the hind wing which are hidden from view by the overlapping fore wing (p. 225).

(19) There is no lack of individual variability among the species of the Danaoid Heliconidae; yet the species as a whole vary but little from the two great types of color-pattern represented by Melinaea and Ithomia. In order to account for this remarkable fact I am forced to resort to Fritz Müller's theory of mimicry (p. 225).

TABLE 1.

Showing the Variations in Color of the "Inner Rufous" (Area 1 in Figures on Plate **4**) of the *fore wing* in the Danaoid Heliconidae.

GENERA.	Rufous.	Translucent rufous	Translucent, slightly rufous	Transparent.	Transparent, slightly yellow	Translucent yellow.	Yellow.	Black.	White.
Lycorea	5								
Ituna				2				1	
Athesis			2		1				
Thyridia				5					
Athyrtis	2								
Olyras	1			1	1			1	
Eutresis		2							
Aprotopos				1				1	
Dircenna	1	3	2	4	1			1	
Callithomia	1							2	
Epithomia	1	1							
Ceratinia	28	1	2	7	1			2	
Sais	5								
Scada						7			
Mechanitis	18					1	1	4	
Napeogenes	7	3	2	12		3		3	
Ithomia	29	13	15	120	26	5		3	1
Aeria						4			
Melinaea	22					2			
Tithorea	4							6	
Total	124	23	23	152	30	22	1	24	1
Excluding Ithomia	95	10	8	32	4	17	1	21	0

Note: The costal edge of the fore wing is usually black; it is rufous or brown, however, in 47 Ithomias and dull yellow in one; it is rufous in two species of Sais, in one species of Ceratinia, and in one species of Athesis. Hence it is black in 348 species and light colored in 52.

TABLE 2.

Showing the Variation (presence or absence) of the "Inner Black" (Area II) of the *fore wing* in the Danaoid Heliconidae.

GENERA.	Present.	Absent.	Remarks.
Lycorea	5		
Ituna	3		
Athesis	2	1	
Thyridia	5		
Athyrtis	2		
Olyras	4		
Eutresis	2		
Aprotopos	2		
Direenna	8	4	
Callithomia	3		
Epithomia	2		
Ceratinia	29	12	
Sais	4	1	
Scada		7	
Mechanitis	23	1	
Napeogenes	13	17	Appears as 2 spots in 1 species.
Ithomia	72	140	Appears as 2 spots in 1 species.
Acria		4	
Melinaea	21	2	Appears as 2 spots in 5 species.
Tithorea	10		
Total . . .	210	190	
Excluding Ithomia	138	50	

TABLE 3.

Showing the Variations in Color of the "Inner Yellow" (Area III) of the *fore wing* in the Danaoid Heliconidae.

GENERA.	Rufous	Translucent rufous	Translucent, slightly rufous	Transparent	Translucent, slightly yellow	Translucent yellow	Yellow	Black	White, generally translucent
Lycorea							5		
Ituna				3					
Athesis			2	1					
Thyridia				5					
Athyrtis	1						1		
Olyras						4			
Eutresis			1	1					
Aprotopos					1		1		
Dircenna		1	2	3	6		1		
Callithomia							3		
Epithomia						1	1		
Ceratinia	16	1	1	7	7	2	6	1	
Sais	2						2	1	
Scada						7			
Mechanitis	11						13		
Napeogenes	2		4	14	1	4	5		
Ithomia	10	11	14	124	23	9	13	1	7
Aeria							4		
Melinaea	14					4	4		2
Tithorea							6	3	1
Total	56	13	24	158	38	31	64	6	10
Excluding Ithomia	46	2	10	34	15	22	51	5	3

TABLE 4.

Showing the presence or absence of the "Middle Black" Mark (Area IV) of the *fore wing* in the Danaoid Heliconidae.

GENERA.	Present.	Absent.	Present, but changed to some color other than black.
Lycorea	5		
Ituna	3		
Athesis	3		
Thyridia	5		
Athyrtis	2		
Olyras	2		
Eutresis	2		
Aprotopos	2		
Dircenna	11	1	
Callithomia	3		
Epithomia	2		
Ceratinia	34	7	
Sais	5		
Scada	7		
Mechanitis	24		
Napeogenes	25	5	
Ithomia	194	6	12 rufous or brown.
Aeria	4		
Melinaea	23	1	
Tithorea	10		
Total	366	20	12
Excluding Ithomia	172	14	

TABLE 5.

Showing the Variation in Color of the "Middle Yellow" Band (Area V) of the *fore wing* in the Danaoid Heliconidae.

GENERA	Rufous.	Translucent rufous.	Translucent, slightly rufous.	Transparent.	Translucent, slightly yellow.	Translucent yellow.	Yellow.	Black.	White, more or less translucent.
Lycorea							5		
Ituna				3					
Athesis			2	1					
Thyridia				5					
Athyrtis	1						1		
Olyras				1		3			
Eutresis				2					
Aprotopos				1			1		
Dircenna			2	3	2	2	3		
Callithomia							3		
Epithomia						1	1		
Ceratinia			2	6	4	5	24		
Sais							5		
Scada						6		1	
Mechanitis	1				1		19		
Napeogenes	2		3	11	4	4	6		
Ithomia	5	6	12	123	24	8	14		20
Aeria							4		
Melinaea	2			3		2	15		2
Tithorea							7	1	2
Total	14	6	24	159	55	31	108	2	24
Excluding Ithomia	9	0	9	36	11	23	94	2	4

TABLE 6.

Showing approximately the Number of Species in which the "Outer Yellow" (Area VI) of the *fore wing* in the Danaoid Heliconidae appears as a separated Marking. It is usually fused with the "Middle Yellow" Area.

GENERA.	Wholly fused with middle yellow.	Partially fused with middle yellow,	Separate.	Absent.
Lycorea		1	4	
Ituna	2	1		
Athesis	3			
Thyridia	5			
Athyrtis			2	
Olyras	1	3		
Eutresis		2		
Aprotopus		2		
Dircenna	7 ?	5 ?		
Callithomia		1	2	
Epithomia		1	1	
Ceratinia	22	16 about	3 about	
Sais			2	3
Scada			4	3
Mechanitis		1	20	3
Napeogenes	6		24 about	
Ithomia	200 about			
Aeria	4			
Melinaea	1		17	6
Tithorea			10	
Total	about 250	about 30	about 90	perhaps 20

TABLE 7.

Showing the Degree of Development of the " Outer Black " (Area VII) of the *fore wing* in the Danaoid Heliconidae.

GENERA.	Well developed over a large area of the fore wing.	Reduced to the outer margin of the fore wing.	Present, but changed to another color.
Lycorea	5		
Ituna		3	
Athesis	2	1	
Thyridia		5	
Athyrtis	2		
Olyras	3	1	
Eutresis	2		
Aprotopos	1	1	
Dircenna	5	7	
Callithomia	3		
Epithomia	2		
Ceratinia	28	13	2 partly rufous.
Sais	2		3 partly rufous.
Scada	3	4	
Mechanitis	22	2	
Napeogenes	26	4	
Ithomia	161	54	16 rufous or brown.
Aeria	4		
Melinaea	24		1 partly rufous.
Tithorea	10		
Total	305	95	22 partly rufous.

TABLE 8.

Showing the presence or absence of the "Longitudinal Black Stripe" (Area VIII) which runs parallel with the lower Edge of the *fore wing* in the Danaoid Heliconidae.

GENERA	Present and well developed as a stripe.	Much reduced.	Absent	Whole area suffused with black.
Lycorea	5			
Ituna	3			
Athesis	3			
Thyridia	5			
Athyrtis	2			
Olyras	3			1
Eutresis	2			
Aprotopos	1			1
Dircenna	7	3		2
Callithomia			1	2
Epithomia.	1	1		
Ceratinia	37	2		2
Sais	3	1	1	
Scada	6		1	
Mechanitis	17		2	5
Napeogenes	20	6		4
Ithomia	200	6	4	2
Aeria	4			
Melinaea	14	5	5	
Tithorea	5			6
Total	338	24	14	24

TABLE 9.

Showing the Manner of Occurrence of the Marginal Spots (Area IX) of the *fore wing* in the Danaoid Heliconidae.

GENERA.	With out spots	1 spot	2 spots	3 spots	4 spots	5 spots	6 spots	7 spots	8 spots	9 spots
Lycorea	5									
Ituna	3									
Athesis	3									
Thyridia	4		1							
Athyrtis	1							1		
Olyras	3						1			
Entresis	2									
Aprotopos	1	1								
Dircenna	11		1							
Callithomia	1			2						
Epithomia	2									
Ceratinia	19	1	3	1			4	12	1	
Sais	5									
Scada	4		1	1				1		
Mechanitis	17			1	2		1	3		
Napeogenes	14			1	3	1	5	3	3	
Ithomia	137	2	14	14	7	16	14	8		
Aeria	4									
Melinaea	13		1			2	4	3	1	
Tithorca	5		1		1			2		1
Total	254	4	22	20	13	19	29	33	5	1

TABLE 10.

Showing the Color-Variations affecting the " Inner Rufous " (Area X) of the *hind wing* in the Danaoid Heliconidae.

GENERA.	Rufous.	Translucent rufous.	Translucent, slightly rufous.	Transparent.	Translucent, slightly yellow.	Translucent yellow.	Yellow.	Black.
Lycorea	5							
Ituna		1		2				
Athesis			2	1				
Thyridia				4	1			
Athyrtis	2							
Olyras	3					1		
Eutresis		2						
Aprotopos	1			1				
Dircenna			8	2	2			
Callithomia	3							
Epithomia	2							
Ceratinia	23		3	6	6	1	2	
Sais	5							
Scada						7		
Mechanitis	16						4	4
Napeogenes	7	6		6	6	5		
Ithomia	31	14	12	133	14	5	1	2
Aeria							4	
Melinaea	19					4	1	
Tithorea	6							4
Total	123	23	25	155	20	23	12	10

TABLE II.

Showing the Variations of the " Middle Black Stripe " (Area XI) of the *hind wing* in the Danaoid Heliconidae.

GENERA.	Present.	Absent.	Fused with the marginal black.	Changed color
Lycorea	5			
Ituna	2	1		
Athesis	2	1		
Thyridia	2	2	1 partially	
Athyrtis	2			
Olyras	1	2		{ 1 changed to translucent yellow ?
Eutresis		2		
Aprotopos	2			
Direnna	4	7	1	
Callithomia			3	
Epithomia		2		
Ceratinia	21		20	
Sais		1	4	
Scada		7		
Mechanitis	22	2		
Napeogenes	15		15	
Ithomia	45	1	168	
Aeria		4		
Melinaea	15	9		
Tithorea	8	2		
Total	146	43	212	1 ?

TABLE 12.

Showing the Color-Variations of the "Outer Rufous" (Area XII) of the *hind wing* in the Danaoid Heliconidae.

GENERA.	Rufous.	Translucent rufous.	Translucent, but slightly rufous.	Transparent.	Translucent, but slightly yellow.	Translucent yellow.	Yellow.	Black.	White, somewhat translucent.
Lycorea	5								
Ituna		1		2					
Athesis			2	1					
Thyridia				4	1				
Athyrtis	2								
Olyras	3					1			
Eutresis	1	1							
Aprotopos	1			1					
Dircenna	2	4	3	3					
Callithomia	1							2	
Epithomia	2								
Ceratinia	29	1	1	7	1			2	
Sais	5								
Scada						7			
Mechanitis	21							3	
Napeogenes	15							15	
Ithomia	44			2	1			165	
Aeria								4	
Melinaea	17						1	6	
Tithorea	5						2	2	1
Total	153	7	6	20	3	8	3	199	1
Excluding Ithomia	109	7	6	18	2	8	3	34	1

TABLE 13.

Showing the presence or absence, and Color-Changes of the "Outer Black" (Area XIII) of the *hind wing* in the Danaoid Heliconidae.

GENERA.	Present.	Absent.	Changed color.
Lycorea	5		
Ituna	3		
Athesis	3		
Thyridia	5		
Athyrtis	2		
Olyras	1		
Eutresis	2		
Aprotopos	2		
Dircenna	12		
Callithomia	3		
Epithomia	2		
Ceratinia	41		
Sais	5		
Scada	7		
Mechanitis	24		
Napeogenes	30		
Ithomia	210	1	11 changed to rufous or brown.
Aeria	4		
Melinaea	24		
Tithorea	10		
Total	398	1	11

TABLE 14.

Showing the Number of the Marginal Spots of the *hind wing* in the Danaoid Heliconidae.

GENERA.	Without spots.	1 spot.	2 spots.	3 spots.	4 spots.	5 spots.	6 spots.	7 spots.	8 spots.	9 spots.
Lycorea		1					3	1		
Ituna	2	1								
Athesis				1	1		1			
Thyridia	4	1								
Athyrtis	1									1
Olyras	3		1							
Eutresis	1						1			
Aprotopos				2						
Dircenna	10		1						1	
Callithomia	2		1							
Epithomia	1					1				
Ceratinia	18	3	4	2	5	5	2	4	1	
Sais	4				1					
Scada	3				1	1	2			
Mechanitis	18		1	1	2	2				
Napeogenes	21		1	1	1	5		1		
Ithomia	164		8	6	12	9	6	6	1	
Acria	4									
Melinaea	19				1			3	1	
Tithorea	4				2	2	2			
Total	279	6	14	13	26	25	17	15	4	1

TABLE 15.

Showing the Color-Variations of the "Inner Rufous" Area of the *fore wing* in Heliconius and Eueides.

	Rufous.	Reddish rufous.	Yellow.	Ocher.	White.	Black.	Iridescent blue
Heliconius	35	8	11		2	29	26
Eueides	14			1		3	
Total	49	8	11	1	2	32	26

Note The costal edge of the fore wing is always black

TABLE 16.

Showing the Variations affecting the "Inner Black" Area of the *fore wing* in Heliconius and Eueides.

	Black.	Iridescent blue.	Rufous.
Heliconius	85	26	
Eueides	13		5
Total	98	26	5

Note: In 54 species of Heliconius the inner rufous is entirely suffused with black.

TABLE 17.

Showing the Color-Variations of the "Inner Yellow" Area of the *fore wing* in Heliconius and Eueides.

	Rufous.	Red.	Yellow.	Ocher.	White.	Black.	Iridescent blue.
Heliconius	11	12	54		20	12	3
Eueides	6			12			
Total	17	12	54	12	20	12	3

TABLE 18.

The " Middle Black " Area in the *fore wing* in Heliconius is present as a Black or Blue Marking in 99 Species and absent in 12. It is present as a Black Mark in all 18 Species of Eueides.

TABLE 19.

Showing the Color-Variation of the " Middle Yellow " Area of the *fore wing* in Heliconius and Eueides.

	Rufous.	Reddish rufous.	Yellow.	Ocher.	White.	Black.
Heliconius	11	12	65		23	
Eueides	5			12	1	
Total	16	12	65	12	24	

TABLE 20.

Showing the Color-Variations of the " Outer Yellow " Area of the *fore wing* in Heliconius and Eueides.

	Rufous.	Reddish rufous.	Yellow.	Ocher.	White.	Black.	Iridescent blue.
Heliconius	2	1	17		24	33	4
Eueides				6	1	11	
Total	2	1	17	6	25	44	4

TABLE 21.

The " Outer Black " Area of the *fore wing* in all the 111 species of Heliconius known to me is Black or Iridescent Blue.

It is Black in all 18 Eueides.

TABLE 22.

Showing the Manner of Occurrence of the Marginal Spots of the *fore wing* in Heliconius and Eueides.

	With-out spots	1 spot.	2 spots.	3 spots.	4 spots	5 spots.	6 spots.	7 spots.	8 spots.	9 spots.
Heliconius	89		5	2	5	3	1	2		1
Eueides	14		3				1			
Total	103		8	2	5	3	5	2		1

TABLE 23.

Showing the Variations affecting the " Longitudinal Black Stripe" of the *fore wing* in Heliconius and Eueides.

	Whole area suffused with black.	Well developed as a black stripe.	Absent (area suffused with rufous).
Heliconius	75	23	13
Eueides	2	16	
Total	77	39	13

TABLE 24.

Showing the Color-Variations of the " Inner Rufous" Area of the *hind wing* in Heliconius and Eueides.

	Rufous.	Reddish rufous	Yellow.	Ochre.	Iridescent green.	Black.	Iridescent blue.	Black and yellow.	Black and rufous.	Black and reddish rufous.
Heliconius	42	7	3		1	16	26	10	1	5
Eueides	15		2			1				
Total	57	7	3	2	1	17	26	10	1	5

TABLE 25.

Showing the Variations of the "Middle Black Stripe" of the *hind wing* in Heliconius and Eueides.

	Well developed as a more or less distinct stripe.	Absent (suffused with black).	Absent (place taken by red or rufous).	Absent (place taken by ocher colors).
Heliconius	47	59	5	
Eueides . .	6		10	1
Total	53	59	15	1

TABLE 26.

Showing the Color-Variations of the "Outer Rufous" Area of the *hind wing* in Heliconius and Eueides.

	Rufous.	Reddish rufous.	Yellow.	White.	Ocher.	Black.	Iridescent blue.
Heliconius .	30	4	19	3		49	6
Eueides . .	17					1	
Total . . .	47	4	19	3		50	6

TABLE 27.

The "Outer Black" Area of the *hind wing* is Black in 106 species of Heliconius, White in 4, and Yellow in 1. It is Black in all the 18 species of Eueides known to me.

TABLE 28.

Showing the Number of Species in each Genus of the Heliconidae examined, and also the Number known according to the Enumeration of Staudinger ('84-'88).

GENERA	Number of species examined by me.	Number of species known to Staudinger ('84-'88).
Lycorea	4 species and 1 var.	4 species and 1 var.
Ituna	3	4
Athesis	3	4
Thyridia	5	4
Athyrtis	2	2
Olyras	4	5
Eutresis	2	2
Aprotopos	2	4
Dircenna	12	20+
Callithomia	3	8
Epithomia	2	2
Ceratinia	41	50+
Sais	5	4
Scada	7	9
Mechanitis	10 species, 14 var.	10 species, 13 var.
Napeogenes	30	30+
Ithomia	242	230+
Aeria	4	4
Melinaea	24	25
Tithorea	10	18
Total of Danaoid Heliconidae.	400	453+
Heliconius	111	130
Eueides	18	24
Total	529	607+

BIBLIOGRAPHY.

Agassiz, A.

'59. [Mechanism of the Flight of Lepidoptera.] Proc. Bost. Soc. Nat. Hist., Vol. 6, p. 426-427.

Bates, H. W.

'62. Contributions to an Insect Fauna of the Amazon Valley. Lepidoptera: Heliconidae. Trans. Linn. Soc. London, Vol. 23, p. 495-566, pl. 55-56.

Bates, H. W.

'63. On a Collection of Butterflies brought by Messrs. Salvin and Godman from Panama, with Remarks on geographical Distribution. Proc. Zool. Soc. London, p. 239-249, pl. 20.

Bates, H. W.

'64-'65. New Species of Butterflies from Guatemala and Panama, collected by Osbert Salvin and F. Du Cane Godman, Esqs. Ent. Mo. Mag., Vol. 1, p. 1-6, 31-35, 55-59, 81-85, 113-116, 126-131, 161-164, 178-180, 202-205. [New Heliconidae, p. 31-35, 55-59.]

Bateson, W.

'94. Materials for the Study of Variation treated with especial Regard to Discontinuity in the Origin of Species. London and New York, 16 + 598 pp. [p. 288-302.]

Beddard, F. E.

'92. Animal Coloration. London and New York, 8 + 288 pp., 4 pls.

Belt, T.

'74. The Naturalist in Nicaragua. London, 403 pp. Second edition. London, 1888, 33 + 403 pp.

Bemmelen, J. F. van.

'89. Ueber die Entwicklung der Farben und Adern auf den Schmetterlings-flügeln. Tijdschrift der Nederlandsche Dierkundige Vereeniging, Ser. 2, Deel 2, p. 235-247.

Blackiston, T., and Alexander, T.

'84. Protection by Mimicry — A Problem in Mathematical Zoology. Nature, Vol. 29, p. 405-406.

Burgess, E.

'80. Contributions to the Anatomy of the Milk-weed Butterfly Danais archippus (Fabr.). Anniversary Mem. Bost. Soc. Nat. Hist., 1880, 16 pp., 2 pls.

Burmeister, H.

'78. Lépidoptères. Description Physique de la République Argentine, Tome 5, 524 pp., 24 pls. [p.21-28.]

Butler, A. G.
  '65. Description of six new Species of Diurnal Lepidoptera in the British
     Museum Collection. Proc. Zool. Soc. London, 1865, p. 430–434, pl. 25.
Butler, A. G.
  '69. Remarks upon certain Caterpillars, etc., which are unpalatable to their
     enemies. Trans. Ent. Soc. London, 1869, p. 27–29. [Feeding habits of
     birds; warning coloration.]
Butler, A. G.
  '69a. Descriptions of several new Species of Nymphalidian Rhopalocera.
     Ann. Mag. Nat. Hist., Ser. 4, Vol. 3, p. 17–21, pl. 9. [New Heliconidae.]
Butler, A. G.
  '69–'74. Lepidoptera Exotica. London, 190 pp., 64 pls. [New Heliconidae.]
Butler, A. G.
  '77. List of Lepidoptera recently collected by Mr. Walter Davis in Peru,
     with Descriptions of a new Genus and several new Species. Ann. Mag.
     Nat. Hist., Ser. 4, Vol. 20, p. 117–129.
Chapman, T. A.
  '88. On Melanism in Lepidoptera. Ent. Mo. Mag., Vol. 25, p. 40.
Coste, F. H. P.
  '90–'91. Contributions to the Chemistry of Insect Colours. Entomologist,
     Vol. 23, p. 128–132, 155–159, 181–187, 217–223, 247–252, 283–287, 309–314,
     338–343, 370–374; Vol. 24, p. 9–15, 37–40, 53–60, 86–94, 114–119, 132–139,
     163–170, 186–192, 206–211. See also Nature, Vol. 45, p. 513–517, 541–542, 605.
Cramer, P.
  1779–'82. Uitlandsche Kapellen. Amsterdam, 4 Vols., 400 pls.
Darwin, C.
  '71. The Descent of Man, and Selection in Relation to Sex. London, 2 Vols.
Dimmock, G.
  '83. The Scales of Coleoptera. Psyche, Vol. 4, p. 3–11, 23–27, 43–47, 63–71
Distant, W. L.
  '82–'86. Rhopalocera Malayana. London, 16 + 481 pp., 44 pls. [p. 33; 129;
     289; 460.]
Dixey, F. A.
  '90. On the phylogenetic Significance of the Wing-markings in certain
     Genera of the Nymphalidae. Trans. Ent. Soc. London, 1890, p. 89–129,
     pl. 1–3.
Druce, H.
  '76. List of the Butterflies of Peru, with Descriptions of new Species. Proc.
     Zool. Soc. London, 1876, p. 205–250, pl. 17–18.
Eimer, G. H. T.
  '89. Die Artbildung und Verwandtschaft bei den Schmetterlingen. Jena;
     12 + 243 pp., 4 Taf.
Felder, C. und R.
  '64–'67. Rhopalocera. Reise Österreichischen Fregatte Novara. Wien,
     6 + 548 pp., 140 Taf. [New Heliconidae.]
Godman, F. D., and Salvin, O.
  '79–'86. Lepidoptera-Rhopalocera. Vol. 1. Biologia Centrali-Americana,
     Insecta. [London], 487 pp., 47 pls. [New Heliconidae.]

Godman, F. D., and Salvin, O.
'80.   A List of Diurnal Lepidoptera collected in the Sierra Nevada of Santa
Marta, Colombia, and the Vicinity. Trans. Ent. Soc. London, 1880, p. 119–
132, pl. 3–4. [New Heliconidae.]

Griffiths, A. B.
'92.   Recherches sur les Couleurs de quelques Insectes. Comptes Rendus
Acad. Sci. Paris, Tome 115, p. 958–959.

Haase, E.
'93.   Untersuchungen über die Mimicry auf Grundlage eines natürlichen
Systems der Papilioniden. Bibliotheca Zoologica, Heft 8, Theil 1, 120 pp.,
6 Taf. [p. 14–15.]

Haase, E.
'93ᵃ. Untersuchungen über die Mimicry auf Grundlage eines natürlichen
Systems der Papilioniden. Bibliotheca Zoologica, Heft 8, Theil 2, 161 pp.,
8 Taf. English translation by C. M. Child. Stuttgart, 1896, 154 pp., 8 pls.

Hewitson, W. C.
'54.   Descriptions of some new Species of Butterflies from South America.
Trans. Ent. Soc. London, New Ser., Vol. 2, p. 245–248, pl. 22–23. [New
species of Heliconidae.]

Hewitson, W. C.
'56-'76. Illustrations of new Species of Exotic Butterflies. London, 5 Vols.,
Plates.

Higgins, H. H.
'68.   On the Colour-Patterns in Butterflies. Quart. Journ. Sci., Vol. 5, p. 325–
329, 1 pl.

Hopkins, F. G.
'89.   Uric Acid Derivatives functioning as Pigments in Butterflies. Proc.
Chem. Soc. London, 1889, p. 117. Also Nature, Vol. 40, p. 335.

Hopkins, F. G.
'91.   Pigment in Yellow Butterflies. Nature, Vol. 45, p. 197.

Hopkins, F. G.
'94.   The Pigments of the Pieridae. A Contribution to the Study of Excre-
tory Substances which function in Ornament. Proc. Roy. Soc. London,
Vol. 57, p. 5–6.

Hopkins, F. G.
'96.   The Pigments of the Pieridae: A Contribution to the Study of Excre-
tory Substances which function in Ornament. Philos. Trans. Roy. Soc.
London, Vol. 186, p. 661–682.

Hübner, J.
'06-'25. Sammlung exotischer Schmetterlinge. Augsburg, 5 Bde., Tafeln.

Humboldt, A. von, et Bonpland, A.
'33.   Recueil d'Observations de Zoologie, Latreille. Paris.

Keeler, C. A.
'93.   Evolution of the Colors of North American Land Birds. Occas. Papers
Cal. Acad. Sci., Vol. 3, 12 + 361 pp., 19 pls.

Kellogg, V. L.
'94.   The Taxonomic Value of the Scales of the Lepidoptera. Kansas Univ.
Quart., Vol. 3, p. 45–89, pl. 9–10.

Kirby, W. F.
  '71-'77. A synonymic Catalogue of Diurnal Lepidoptera. London, 5 + 690
    pp. Supplement. London, p. 691-883.
Leydig, F.
  '55. Zum feineren Bau der Arthropoden. Müller's Archiv, Jahrg. 1855
    p. 376-380, Taf. 15-18.
Mayer, A. G.
  '96. The Development of the Wing Scales and their Pigment in Butterflies
    and Moths. Bull. Mus. Comp. Zoöl. Harv. Coll., Vol. 29, p. 209-236, 7 pls.
Ménétriés, E.
  '63. Descriptions des nouvelles Espèces de Lépidoptères. St. Petersbourg.
    [New Heliconidae.]
Merrifield, F.
  '94. Temperature Experiments in 1893 on several Species of Vanessa and
    other Lepidoptera. Trans. Ent. Soc. London, 1894, p. 425-438, pl. 9.
Moore, F.
  '90-'96. Lepidoptera Indica. London, 2 Vols., 190 pls.
Müller, F.
  '78. Notes on Brazilian Entomology. Trans. Ent. Soc. London, 1878, p. 211-
    223. [Odors emitted by Butterflies and Moths.]
Müller, F.
  '79. Ituna and Thyridia. Ein merkwürdiges Beispiel von Mimicry bei
    Schmetterlingen. Kosmos, Bd. 5, p. 100-108, fig. 1-4. English translation
    by Raphael Meldola. Trans. Ent. Soc. London, 1879, p. 20-29 Proc., fig. 1-4.
Müller, W.
  '86. Südamerikanische Nymphalidenraupen. Versuch eines natürlichen
    Systems der Nymphaliden. Zool. Jahrbücher, Bd. 1, p. 417-678, Taf.
    12-15.
Poulton, E. B.
  '85. The essential Nature of the Colouring of Phytophagous Larvae (and
    their Pupae); with an Account of some Experiments upon the Relation
    between the Colour of such Larvae and their Food-plant. Proc. Roy. Soc.
    London, Vol. 38, p. 269-315.
Poulton, E. B.
  '87. The experimental Proof of the Protective Value of Colour and Markings
    in Insects in Reference to their Vertebrate Enemies. Proc. Zool. Soc.
    London, 1887, p. 191-274.
Poulton, E. B.
  '90. The Colours of Animals. International Sci. Series, 67. New York,
    13 + 360 pp.
Poulton, E. B.
  '93. The experimental Proof that the Colours of certain Lepidopterous
    Larvae are largely due to modified Plant Pigments derived from Food.
    Proc. Roy. Soc. London, 1893, p. 417-430, pl. 3-4.
Ridgway, R.
  '86. A Nomenclature of Colors for Naturalists. Boston, 129 pp., 17 pls.
Rippon, R. H. F.
  '89-'96. Icones Ornithopterorum: A Monograph of the Rhopalocerous Genus
    Ornithoptera, or Bird-wing Butterflies. London, parts 1-5, Plates.

Schäffer, C.

'89. Beiträge zur Histologie der Insekten. Zool. Jahrbücher, Morph. Abth.,
Bd. 3, p. 611-652, Taf. 29-30. [Reference, p. 647-652.]

Schatz, E., und Röber, J.

'85-'92. Die Familien und Gattungen der Tagfalter. (Exotische Schmetterlinge von Standinger und Schatz.) Fürth, 2. Theil, 2 + 284 pp., 50 Taf.

Scudder, S. H.

'88-'89. The Butterflies of the Eastern United States and Canada with special
Reference to New England. Cambridge, 3 Vols., 24 + 1958 pp., 89 pls.

Seitz, A.

'89. Lepidopterologische Studien im Ausland. Zool. Jahrbücher, Syst. Abth.,
Bd. 4, 771-779, 905-924.

Semper, G.

'86-'92. Die Schmetterlinge der Philippinischen Inseln. Wiesbaden, 2 Bde.

Snellen, P. C. T., en Leeuwen, J. van, Jr.

'87. Bijdrage tot de Kennis der Lepidoptera van het Eiland Curaçao.
Tijdsch. voor Ent., Deel 30, p. 9-66, pl. 1-5.

South, R.

'89. Notes on some Aberrations in the Genus Vanessa. Entomologist,
Vol. 22, p. 217-224, pl. 8. [Fig. 7.]

Sruka, A.

'84. Eine neue Athyrtis. Lepidoptera: Fam. Heliconidae. Berlin. Ent.
Zeitschr., Bd. 28, p. 163-165.

Sruka, A.

'85. Neue Südamerikanische Danaidae und Heliconidae. Berlin. Ent.
Zeitschr., Bd. 29, p. 121-130, Taf. 1.

Standinger, O.

'82. On three new and interesting Species of Rhopalocera. Proc. Zool. Soc.
London, 1882, p. 396-398, pl. 24. [New Heliconidae.]

Standinger, O.

'84-'88. Exotische Tagfalter. (Exotische Schmetterlinge von Standinger
und Schatz.) Fürth, 1. Theil, 333 pp., 100 Taf. [New Heliconidae.]

Urech, F.

'91. Beobachtungen über die verschiedenen Schuppenfarben und die zeit-
liche Succession ihres Auftretens (Farbenfelderung) auf den Puppen-
flügelchen von Vanessa urticae und io. Zool. Anzeiger, Jahrg. 14, p. 466-
473.

Urech, F.

'92. Über Eigenschaften der Schuppenpigmente einiger Lepidopteren-
Species. Zool. Anzeiger, Jahrg. 15, p. 299-306.

Urech, F.

'93. Beiträge zur Kenntniss der Farbe von Insektenschuppen. Zeitschr. f
wiss. Zool., Bd. 57, p. 306-384.

Van Bemmelen. See Bemmelen. J. F. van.

Wallace, A. R.

'67. [Theory of Warning Coloration.] Trans. Ent. Soc. London, Ser. 3,
Vol. 5, p. 80-84, Proc.

Wallace, A. R.

'89. Darwinism. London and New York, 16 + 494 pp. [p. 253-267.]

Walsingham.

'85. On some probable Causes of a Tendency to melanic Variation in Lepidoptera of high Latitudes. Trans. Yorkshire Union, pt. 5, p. 113-140. Cf. Nature, Vol. 31, p. 505.

Watkins, W.

'91. Ornithoptera trojana, Staudinger. Entomologist, Vol. 24, p. 178-179, pl. 4.

Weismann, A.

'82. Studies in the Theory of Descent. Translated and edited by Raphael Meldola. London, 2 Vols., 28 + 720 pp., 8 pls. [p. 1-158.]

Weymer, G.

'75. Exotische Lepidopteren. Stettiner Ent. Zeit., Jahrg. 36, p. 368-385, Taf. 1-2.

Weymer, G.

'84. Exotische Lepidopteren. II. Stettiner Ent. Zeit., Jahrg. 45, p. 7-28, Taf. 1-2.

TABLE OF CONTENTS.

PART A.

GENERAL PHENOMENA OF COLOR IN LEPIDOPTERA.

I. CLASSIFICATION OF COLORS.

II. THE ESSENTIAL NATURE OF PIGMENTAL COLOR IN LEPIDOPTERA.

III. DEVELOPMENT OF THE VARIOUS COLORS IN THE PUPAL WINGS.

IV. THE LAWS WHICH GOVERN THE COLOR-PATTERNS OF BUTTERFLIES AND MOTHS.

V. THE CAUSES WHICH HAVE LED TO THE DEVELOPMENT AND PRESERVATION OF THE SCALES OF THE LEPIDOPTERA.

PART B.

COLOR VARIATIONS IN THE HELICONIDAE.

I. GENERAL CAUSES WHICH DETERMINE COLORATION IN THE HELICONIDAE.

II. METHODS PURSUED IN STUDYING THE COLOR-PATTERNS OF THE HELICONIDAE.

PART C.

PLATE I.

ABBREVIATIONS.

B. Back surface covered with wings.   O. Orifice for admission of light.
F. Front surface covered with wings.   S. Spectroscope.

Arrow indicates directions of rays of light.

Fig. 1.   Perspective view of spectroscopic apparatus used in determining the composition of the colors of Lepidoptera.

Fig. 2.   Horizontal section of same.   See p. 175.

Fig. 3.   Pendulum used in determination of the frictional resistance between the air and the wings of Lepidoptera.   See p. 193.

Figs. 4, 5.   Diagrams to illustrate Keeler's method of projection, as applied to Lepidoptera.   See p. 207.

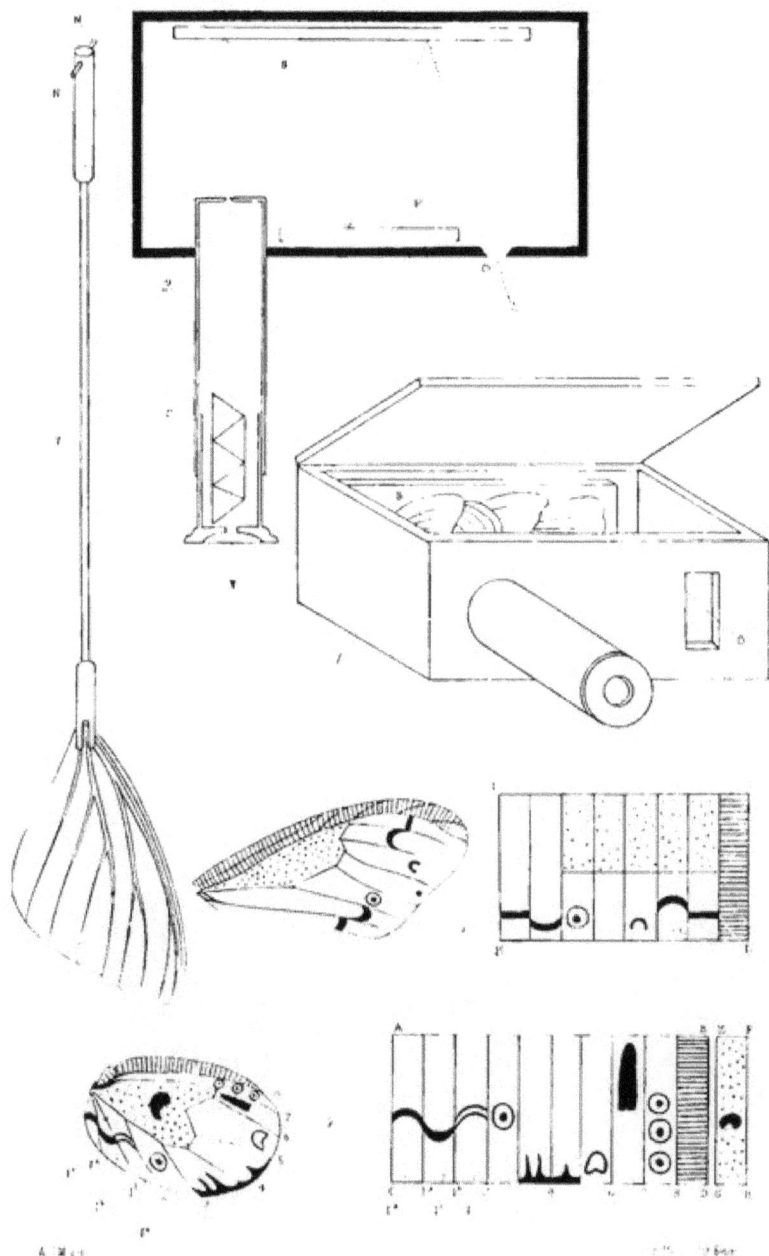

PLATE 2.

Diagrams to illustrate the laws which govern the Color-Patterns of Lepidoptera.

Fig. 6. Euthalia bellata (W. L. Distant, '82–'86, Plate 43, Fig. 12). Illustrates the law of bilaterality of spots. See p. 183.

Fig. 7. Zethera musa (G. Semper, '86–'92, Taf. 7, Fig. 10). Bilaterality of double spots. See p. 183.

Fig. 8. Eye-spots in Morpho. See p. 182, 183.

Fig. 9. Parthenos gambrisius (W. L. Distant, '82–'86, Plate 11, Fig. 7). A series of complex spots, each one being similar to the rest, and bilaterally symmetrical.

Figs. 10, 11. Ornithoptera urvillana and O. priamus (R. H. F. Ripon, '89–'93). Spots within spots, all being bilaterally symmetrical.

Figs. 12, 13. Hestia jasonia and H. leuconoe. Axis of lateral symmetry (H, H), for spots passes through center of interspace. H. jasonia (F. Moore, '90–'96, Plate 3, Fig. 1). H. leuconoe (G. Semper, '86–'92, Taf. 1, Fig. 3).

Fig. 14. Papilio emalthion, to illustrate fusion of two rows of spots.

Fig. 15. Ornithoptera trojana, an apparent exception to the law of bilaterality. See p. 187.

Fig. 16. Limenitis proserpina (S. H. Scudder, '88–'89, Plate 2, Fig. 9), showing fusion of two rows of differently colored spots. See p. 187.

Fig. 17. Saturnia spini, false eye-spot. See p. 187.

Fig. 18. Cases of degeneration of bands of color. See p. 184.

Fig. 19. Missanga patina (F. Moore, '90–'96, Plate 72, Fig. 2c). Exceptional form of eye-spot. See foot note p. 186.

Figs. 20–23. Hypothetical conditions of coloration, not found in nature, being contrary to the laws of color-pattern. See p. 188.

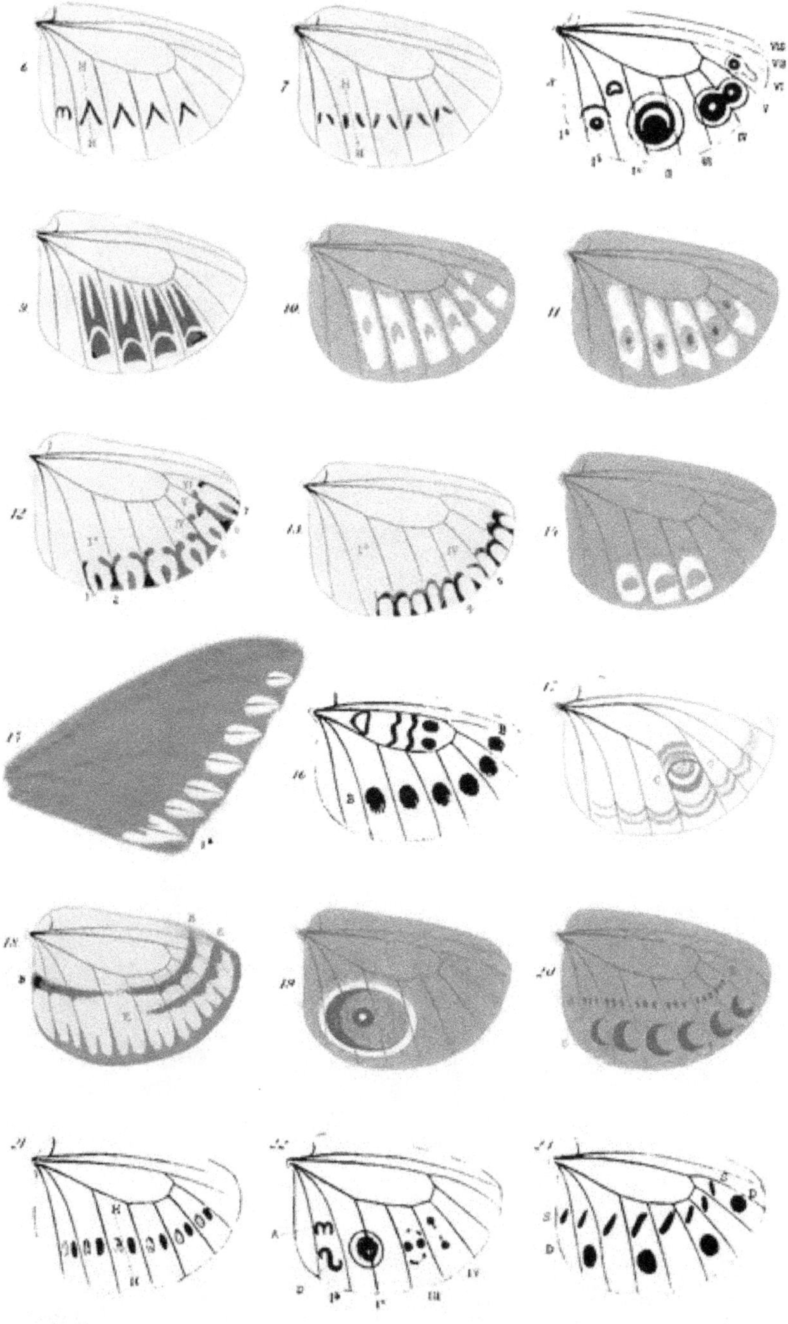

To illustrate color-development in Callosamia promethea and Danais plexippus.

Fig. 24.   Enlarged view of pupal wing of C. promethea in the "white stage."
   See p. 178.

Fig. 25.   Scale from wing of C. promethea in white stage of color development,
   showing the total absence of pigment in the scale.   See p. 178.

Fig. 26.   Scale from light drab-colored area of mature wing of C. promethea.

Figs. 27, 36, and 33.   Successive stages in the formation of color in pupal hind
   wing of C. promethea.

Figs. 28, 37–40.   Successive stages in the formation of color in the pupal fore
   wing of ♀ C. promethea.   See p. 179, 180.

Figs. 29, 30–35.   Successive stages in the formation of color in the pupal wings
   of ♂ C. promethea.   (Figs. 29, 30–33, 35 fore wing; Fig. 31 hind
   wing.)   See p. 179, 180.

Figs. 34 and 41.   Pupal hind wings of C. promethea, respectively mature ♂
   and ♀.

Figs. 42–45.   Successive stages in the color-development of D. plexippus.   See
   p. 180–181.

PLATE 4.

Systematic analysis of the characteristic markings upon the wings of the Heli-coniidae. Homologous markings are designated by the same numerals, I, II, III, etc.

Fig. 46. Lycorea ceres; an example of the "Melinaea type" of coloration. See p. 205.

Fig. 47. Thyridia psidii; an example of the "Ithomia type" of coloration. See p. 206 and Plate 7, Fig. 79.

Fig. 48. Melinaea paraiya.

Fig. 49. Ceratinia ninonia.

Fig. 50. Heliconius antiochus.

Fig. 51. Napeogenes duessa.

Fig. 52. Ithomia sao.

Fig. 53. Melinaea gazoria.

Fig. 54. Ithomia nise.

Fig. 55. Mechanitis polymnia.

Fig. 56. Eueides eleobaea.

Fig. 57. Tithorea furia var.

Fig. 58. Heliconius eucrate.

Fig. 59. Heliconius melpomene.

Fig. 60. Heliconius erato.

PLATE 4.

MAYER - COLOR AND COLOR PATTERNS.

PLATE 5.

Color-patterns of the Antiochus and Erato groups of the Heliconidae projected by Keeler's method. See p. 207.

Figs. 61, 62.   Heliconius sara and H. antiochus, to show variation of yellow to white.   See p. 210, and Fig. 50, Plate **4**.

Fig. 63.   H. galanthus, showing development of white.

Fig. 64.   H. charitonia, rows of double spots.

Fig. 65.   H. phyllis, close relation between yellow and red.

Fig. 66.   H. ricini.

Figs. 67, 68.   H. erato; two color-types.

Fig. 69.   H. claudia; an example of the Sylvanus group.

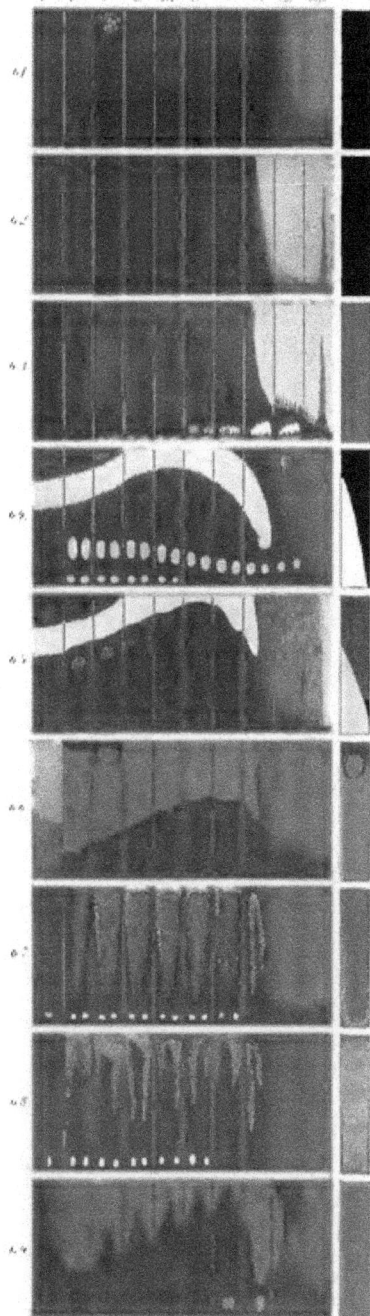

i

PLATE 6.

Color-patterns of the Melpomene group of Heliconius and of the genus Eucides.

Fig. 70. H. melpomene, the type of the Melpomene group. See p. 212, and Fig. 50, Plate 4.

Fig. 71. H. melpomene var. callieopis, showing the breaking up of the red area of the primaries.

Fig. 72. H. melpomene var. cybele; the fore wing has assumed a color-pattern which recalls the "Melinaea type" of coloration found in the Danaoid Heliconidae.

Fig. 73. H. thelxiope, derived phylogenetically from H. melpomene, and showing a rather close approach to the "Melinaea type" of coloration. See p. 212.

Fig. 74. H. vesta.

Fig. 75. Eucides thales ♂; represented to show the close resemblance of its color pattern to H. vesta. See p. 212.

Figs. 76, 77. E. mercaui and E. aliphera. E. mercaui is intermediate in color-pattern between E. thales and E. aliphera.

Fig. 78. Eucides cleobaea, to show the close approach of this insect to the "Melinaea type" of coloration.

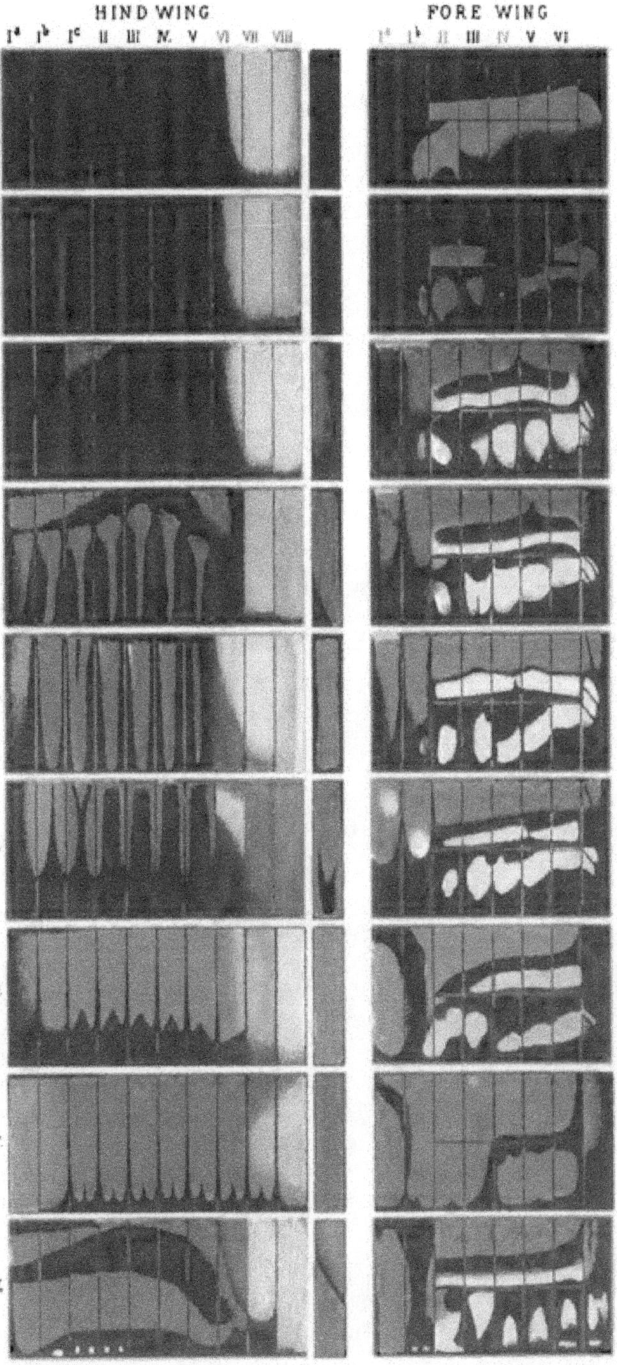

70. *Heliconius melpomene* Linn.

71. *H. melpomene* var. *callycopis* Cram.

72. *H. melpomene* var. *cybele* Cram.

73. *H. Thelxiope* Hübn. ♂

74. *H. vesta* Cram.

75. *Eueides thales* Cram. ♂

76. *E. mereau* Hübn.

77. *E. aliphera* Godt.

78. *E. cleobaea* Hübn.

A. G. M. del. B. Meisel Lith. Boston.

PLATE 7.

Intended to show some types of coloration which are found in the Danaoid Heliconidae, and also the remarkable individual variation in Mechanitis isthmia.

Fig. 79. Thyridia psidii, an example of the "Ithomia" type of coloration. See p. 213, and Plate 4, Fig. 47.

Fig. 80. Napeogenes cyrianassa, showing semi-translucent condition of wings. See p. 213.

Fig. 81. Ceratinia vallonia.

Fig. 82. Melinaea parallelis; albinism of spots on primaries; black band of hind wing broken in the middle. See p. 188, 213.

Fig. 83. Ceratinia leucania, which probably mimics M. parallelis.

Figs. 84-87. Mechanitis isthmia, showing remarkable individual variation in the black stripe of the hind wings, and also in the "inner yellow" spot of the fore wings. See p. 184, 213.

HIND WING.   FORE WING

I^a I^b I^c II III IV V VI VII VIII   I^a I^b II III IV V VI

79. Thyridia
psidii Linn.

80. Napeogenes
cyranassa
Doubl. & Hew.

81. Ceratinia
vallonia Hew.

82. Melinaea parallelis
Butl.

83. Ceratinia leucania
Bates.

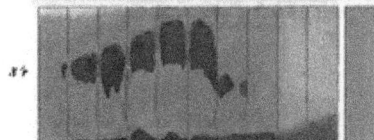

84. Mechanitis
isthmia Bates.

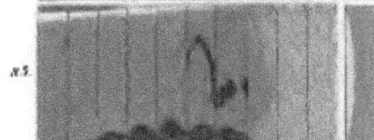

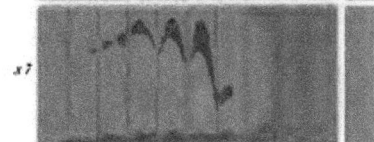

85. Mechanitis
isthmia Bates.

86. Mechanitis
isthmia Bates.

87. Mechanitis
isthmia Bates.

A. G.M. del.   B. Meisel lith. Boston

PLATE 8.

Illustrates the mimicry between members of the "Sylvanus" group of the genus Heliconius and various Melinaeas, etc.

Fig. 88. Heliconius eucoma, an example of the "Sylvanus" type of coloration in genus Heliconius. See p. 214.

Figs. 89, 90. Heliconius dryalus and Melinaea paraiya; close resemblance of their color-patterns. See p. 214.

Figs. 91, 92, 93, 94. Respectively Heliconius eucrate, Melinaea thera, Eueides dianasa, and Mechanitis polymnia; showing close resemblance between color-patterns. See p. 214.

Figs. 95, 96. Heliconius sylvana and Melinaea egina; these two forms are said by Bates to mimic each other. See p. 214.

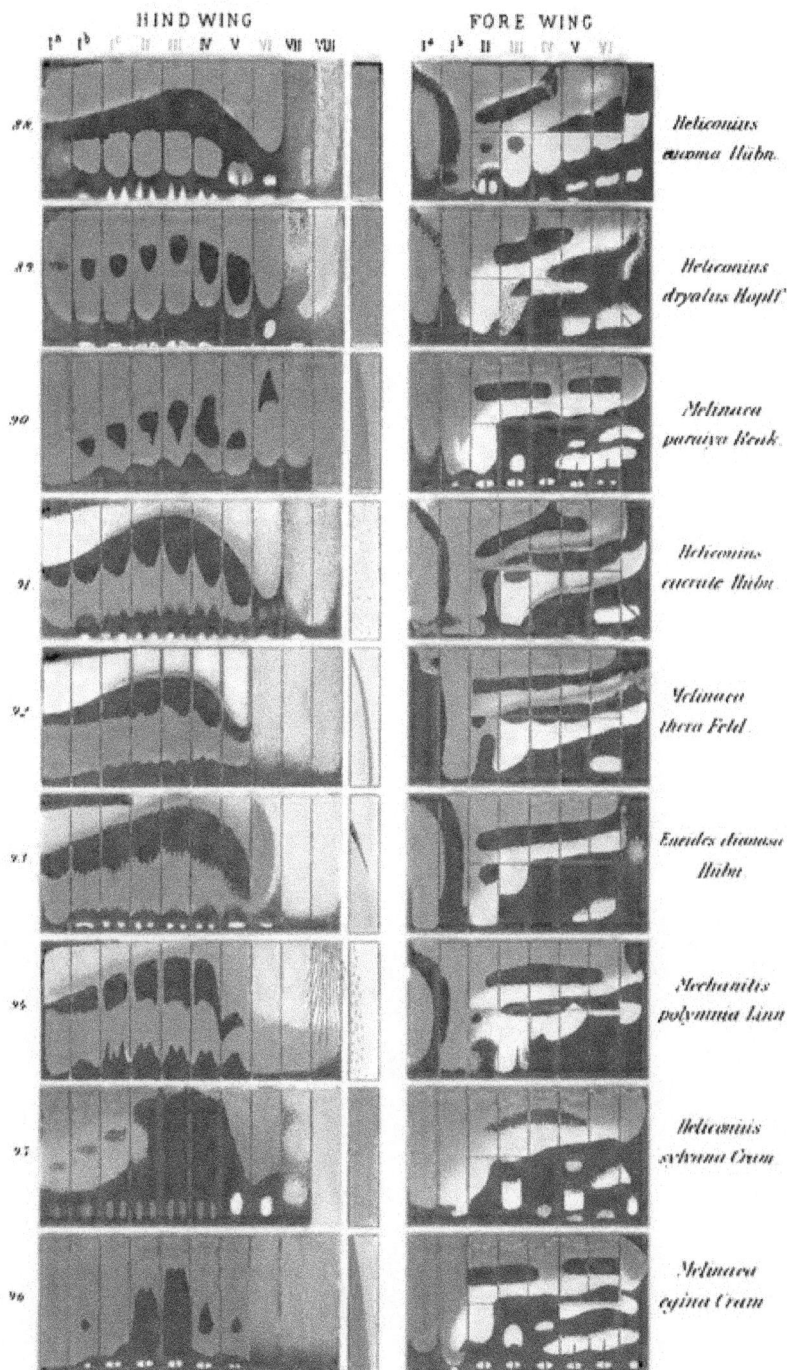

HIND WING

1ᵃ 1ᵇ 1ᶜ II III IV V VI VII VIII

FORE WING

1ᵃ 1ᵇ II III IV V VI

88.   Heliconius euxoma Hubn.

89.   Heliconius dryalus Hopff.

90.   Melinaea paraiya Reak.

91.   Heliconius eucrate Hubn.

92.   Melinaea thera Feld.

93.   Eueides dianasa Hubn.

94.   Mechanitis polymnia Linn.

95.   Heliconius sylvana Cram.

96.   Melinaea egina Cram.

PLATE 9.

Diagrams to illustrate color-variations.

The various colors are laid off at definite intervals along the axis of abscissae, and the ordinates represent the number of species which exhibit the various colors.

Fig. 97. Represents the color variations of the " inner rufous " area of the fore and hind wings in the Danaoid Heliconidae. The full line represents the variations of the *fore* wing. The dotted line those of the *hind* wing. The closeness of these two lines shows the intimate relations between the color-variations of the " inner rufous " areas upon fore and hind wing. See p. 218.

Fig. 98. The full line represents color-variations of " inner yellow " spot of fore wings in Danaoid Heliconidae. The dotted line represents same for " middle yellow." It is apparent that the " inner yellow " is more variable than the " outer yellow," and also that the variations of both are quite similar to those of the " inner rufous." See p. 219.

Fig. 99. Color-variations of " inner rufous " areas of Acraeoid Heliconidae. The full line represents the fore wing and the dotted line the hind wing. See p. 218.

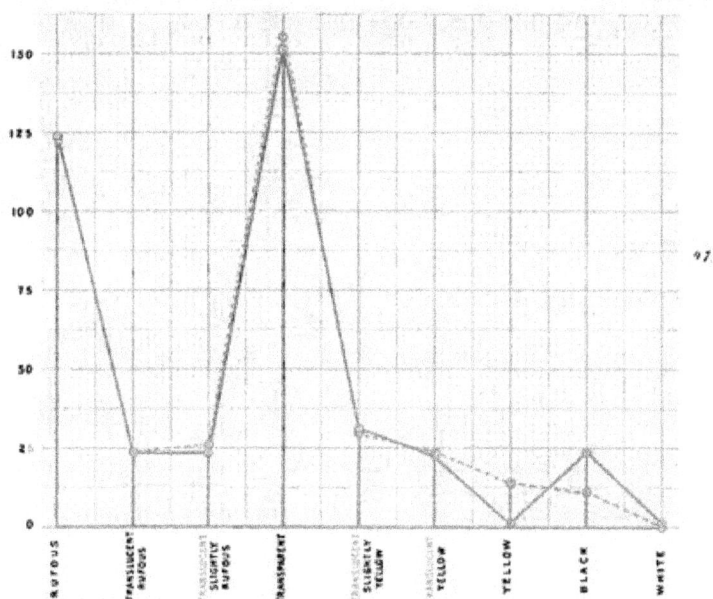

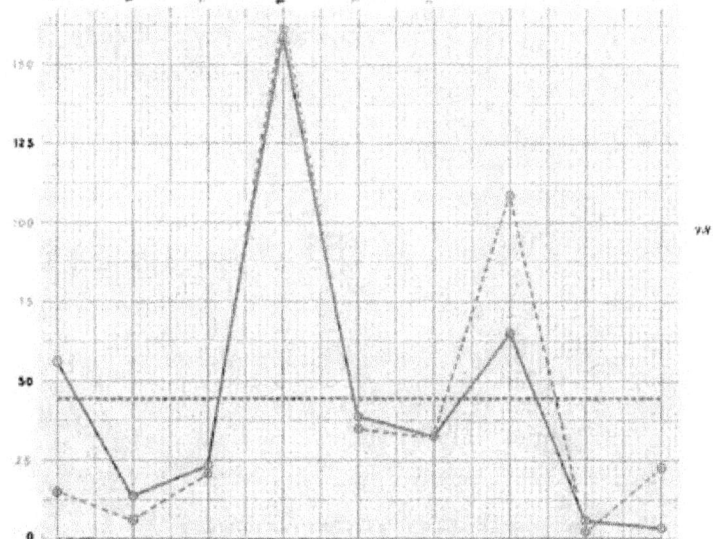

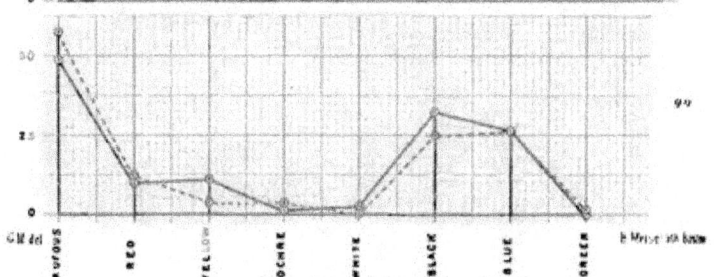

PLATE 10.

Fig. 100. Color-variations of "inner yellow" spots on fore wings of Acraeoid Heliconidae. The full line represents the "inner yellow," the dotted line the "middle yellow." See p. 221.

Fig. 101. Variations of marginal spots upon fore wing in Danaoid Heliconidae. These spots tend to appear either as 2 or 3, or as 6 or 7 spots. See p. 223.

Fig. 102. Variations of marginal spots of hind wings in Danaoid Heliconidae. These spots tend to appear either as 4 or as 5 spots. See p. 223.

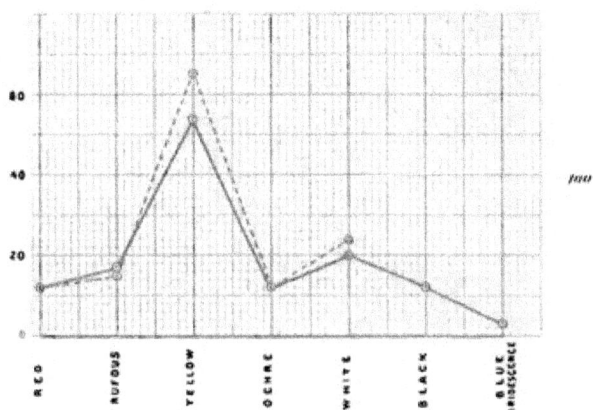

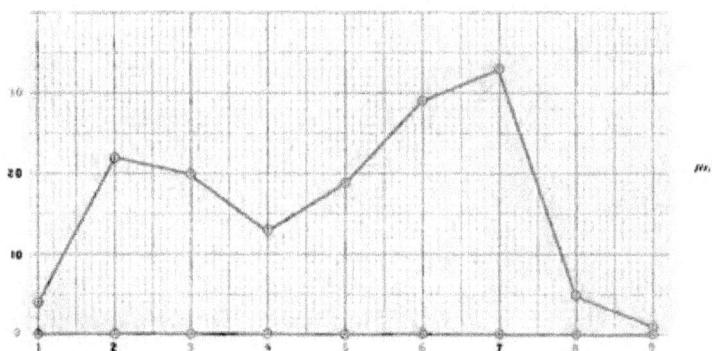

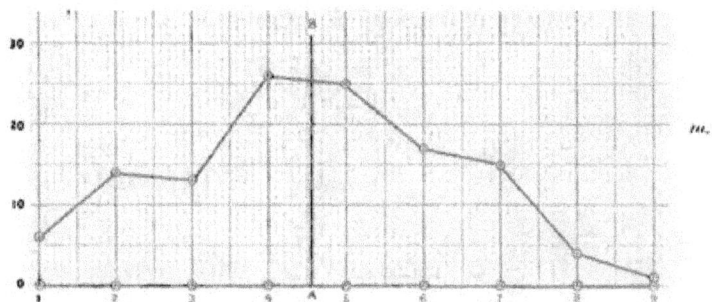

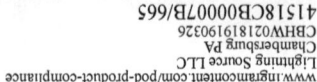